PHYSICAL GEOLOGY LABORATORY MANUAL

# PHYSICAL GEOLOGY
## LABORATORY MANUAL

## A Guide to the Study of
## Earth Materials and Landforms

### R. D. DALLMEYER
### University of Georgia

Murchison, 1867

KENDALL/HUNT PUBLISHING COMPANY
Dubuque, Iowa

Copyright © 1975 by R. David Dallmeyer

ISBN 0—8403—1231—8

Printed in the United States of America

# Contents

# Preface

The exercises in this manual are designed to develop skills in the observation and recognition of geologic features. Emphasis is placed on the deductive processes of observation and not on the memorization of specific categories of data. I hope the manual will foster an awareness of the physical world and the processes which have helped to form it.

I sincerely thank members of the faculty and many graduate students of the Department of Geology at the University of Georgia for their review of preliminary copies of the manuscript. Cover figure from Harper's Magazine, September, 1879.

R.D. Dallmeyer

**The Geologic Time Scale**

| Era | Period* | Epoch | Approximate Ages-$10^6$ yr. |
|---|---|---|---|
| Cenozoic | Quaternary (Q) | Recent<br>Pleistocene | |
| | Tertiary (T) | Pliocene<br>Miocene<br>Oligocene<br>Eocene<br>Paleocene | 1 |
| Mesozoic | Cretaceous (K) | Upper<br>Lower | 65<br>135 |
| | Jurassic (J) | Upper<br>Middle<br>Lower | |
| | Triassic (T̲R̲) | Upper<br>Middle<br>Lower | 180 |
| Paleozoic | Permian (P) | Upper<br>Lower | 225 |
| | Pennsylvanian (IP) | Upper<br>Middle<br>Lower | 275 |
| | Mississippian (M) | Upper<br>Lower | |
| | Devonian (D) | Upper<br>Middle<br>Lower | 350 |
| | Silurian (S) | Upper<br>Middle<br>Lower | 400 |
| | Ordovician (O) | Upper<br>Middle<br>Lower | 435 |
| | Cambrian (C) | Upper<br>Middle<br>Lower | 500 |
| Precambrian (PC) | Locally divided into<br>Early, Middle and Late | | 600<br>3900+ |

*Symbols in parenthesis are abbreviations commonly used to designate the age of rock units on maps.

# Part 1

# MINERALS AND ROCKS
## THE EARTH MATERIALS

Dana, 1937

# Exercise 1

# Introduction to Minerals

Rocks record the basic information which enables us to reconstruct the geologic evolution of the earth. Although there are many different types of rocks, most are made of minerals. Thus, a prerequisite for geologic study is the ability to identify minerals which are commonly found in rocks.

To be a mineral, a material must:

1.  Be naturally occurring.
2.  Be inorganic.
3.  Be a crystalline element or compound.
4.  Have a unique or only limited range in chemical composition.

Because minerals are crystalline solids, they have a specific internal structure and their constituent atoms or ions arranged in an ordered, three-dimensional network. This precise arrangement of atoms or ions extends throughout mineral structures and controls morphology (shape) and physical properties (such as hardness and density). Because most physical properties are relatively constant in all specimens of the same mineral species (regardless of size), they may be used to identify unknown mineral samples.

Before observing physical properties for identification purposes, it is first necessary to determine if a specimen is an individual crystal, a broken fragment of a larger crystal, or an aggregate of small crystals. This distinction is best accomplished by slowly rotating the specimen and observing its reflection of light. If there are a number of small reflecting surfaces, the specimen is probably a crystal aggregate (each crystal reflecting light from its faces), and caution must be used when observing certain physical properties. If one or only a few reflective surfaces are seen, the specimen is either an individual crystal or a fragment of a larger crystal.

Most common rock-forming minerals may be identified by their visible physical properties. The more important properties which are especially helpful in mineral identification are listed below.

1.  *Crystal Form.* If a mineral forms in an environment where its growth is unimpeded (as in open spaces), it may develop smooth crystal faces (Fig. 1-1). Crystal faces reflect the internal atomic order of minerals and are arranged in different geometric shapes, one of which is characteristic for a specific mineral species (regardless of crystal size). However, use of crystal shape for mineral identification is restricted because of the relative scarcity of well-formed crystals.

**Figure 1-1.** Photograph of intergrown fluorite crystals. Note well-developed cubic crystal forms.

2.  *Color.* Color is the most obvious physical property of minerals and for some species it provides a useful clue for identification (for example; pyrite, galena and olivine). However, the color of many mineral species (such as quartz) varies widely and reliance on color as a basis for identification can be misleading. Surface weathering usually modifies the coloration of minerals, so always observe color on fresh, unweathered surfaces.

3.  *Streak.* The color of a powdered portion of a mineral is generally more diagnostic for identification purposes than is overall coloration. The powder may be easily obtained by rub-

bing the corner of a specimen against an unpolished, white porcelain surface (streak plate). For some mineral species, streak color is different than overall coloration. Unfortunately, many light-colored minerals have white streaks which are of little diagnostic value in identification. Also, minerals which are harder than the streak plate will not be powdered during a streak test.

4. *Luster.* The manner in which light is reflected by a mineral surface is termed luster. Two broad categories of luster are generally recognized—metallic (reflection like metal) and nonmetallic. The luster of nonmetallic minerals may be further characterized using descriptive terms such as vitreous (shiny like glass), pearly (reflectance similar to pearl), etc. A list of common nonmetallic lusters and representative mineral examples is presented in Table 1-1. Weathering may affect the luster of a mineral specimen, so be certain to examine fresh surfaces.

**TABLE 1-1**
Varieties of Nonmetallic Luster

| Type | Mineral Examples |
| --- | --- |
| Vitreous (glassy) | Quartz |
| Adamantine (brilliant) | Diamond |
| Resinous | Sphalerite |
| Dull | Limonite, Kaolinite |
| Pearly | Talc, Gypsum |
| Greasy | Graphite |

5. *Diaphaneity.* The amount of light which passes through a specimen is a measure of its diaphaneity. This physical property is described as follows: *transparent*—objects may be seen through specimens (some calcite); *translucent*—objects cannot be seen through specimens, but light is transmitted (look along edges of specimens [gypsum]); *opaque*—light is not transmitted through specimens (all minerals with metallic luster are generally opaque).

6. *Hardness.* A mineral's resistance to abrasion is termed its hardness. A mineral's hardness is commonly measured by how easily it scratches or is scratched by a material of known hardness. The most common standard is the Mohs' scale of scratch hardness (Table 1-2). This scale consists of ten relatively common minerals of different hardness, each assigned a number from 1 to 10 (10 is the hardest and most difficult to scratch). As each mineral on the Mohs' hardness scale has an assigned number, the scale is relative. Thus, fluorite (number 4 on the scale) is not twice as hard as gypsum (number 2 on the scale), nor is it half as hard as topaz (number 8 on the scale). Other convenient materials of known hardness (fingernail, penny, glass, knife blade) may be used to test an unknown mineral's hardness. These are compared to the Mohs' hardness scale in Table 1-2.

**TABLE 1-2**
Mohs' Mineral Hardness Scale

| Material | Hardness | Mohs' Number | Mineral |
| --- | --- | --- | --- |
| | | 1 | talc |
| | | 2 | gypsum |
| Fingernail | 2½ | | |
| | | 3 | calcite |
| Penny | 3½ | | |
| | | 4 | flourite |
| Glass plate | 5-5½ | 5 | apatite |
| Knife blade | 5½-6 | 6 | orthoclase |
| Quartz | 7 | 7 | quartz |
| | | 8 | topaz |
| | | 9 | corundum |
| | | 10 | diamond |

To determine the hardness of a specimen, choose a small, flat surface and attempt to scratch the mineral with a reference material of known hardness (fingernail, penny, glass, knife blade, quartz crystal). If the mineral is scratched, its hardness is less than that of the material used. Check your results by attempting to scratch the reference material with the mineral. It is often difficult to determine whether the mineral or reference material has been scratched. Therefore, check any marks carefully to make certain a scratch does exist. Choose a different hardness standard and repeat the test until the hardness of the unknown mineral is bracketed. You can approximate the hardness between reference materials by estimating the ease with which the unknown mineral or reference material was scratched. A few minerals show variable hardness depending on which direction they are scratched (kyanite), so be certain to try several directions of scratching to determine hardness. Specimens which are aggregates of crystals are difficult to test for hardness because they may be pulled apart without actually testing the hardness of individual crystals.

7. *Cleavage.* Cleavage is the tendency of some minerals to break along zones of weakness in their atomic structures. This orderly breakage results in sets of closely-spaced, smooth parallel surfaces which are termed cleavage planes. Some minerals cleave in only one direction and occur as thin sheets with ragged borders (biotite and muscovite). Others have more than one cleavage direction (Fig. 1-2). The angles between different sets of cleavage planes is consistent for a specific mineral species. The smoothness (or perfection of development) of different sets of cleavage planes may be variable within a specific mineral species. When describing a mineral's cleavage, always be certain to note:

   a. The number of cleavage planes.
   b. The angles between cleavage planes.
   c. The smoothness (or perfection) of the different cleavage planes.

In well-formed mineral specimens, smooth crystal faces may often be confused with cleavage planes. However, crystal faces are only surface features and do not extend throughout mineral structures. Minerals will not break along directions which parallel crystal faces, but will break along sets of cleavage planes. Determination of cleavage

**Figure 1-2.** Photograph of galena crystal showing three sets of cleavage planes at right angles.

properties for crystal aggregates is difficult. Caution must be used so that only cleavages displayed by individual crystals are described.

8. *Fracture.* Some minerals have no systematic zones of weakness within their internal atomic arrangement and break along random zones. The nature of these fracture surfaces may be useful in classifying certain mineral species. The most diagnostic types of fracture are *conchoidal* (concentric, dish-shaped fractures similar to the broken surface of glass) and *fibrous* (splintery breakage).

9. *Specific Gravity.* Specific gravity is a measure of relative weight. It is defined by comparing the weight of a material to the weight of an equal volume of water. Thus, a material with a specific gravity of 3.0 would be three times heavier than an equal volume of water. Accurate determination of specific gravity requires specialized equipment, but relative differences in specific gravity may be determined by comparing the "heft" of a specimen against that of a known mineral species of similar size.

10. *Special Physical Properties.* In addition to the nine major physical properties listed above, there are special physical properties which are particularly useful in identifying minerals which possess them. These include:

   a. *Magnetism.* Some minerals are magnetic and are attracted by a magnet (magnetite).
   b. *Taste.* The taste of some minerals is very diagnostic (halite).
   c. *Odor.* The odor of some minerals (particularly when moist) is characteristic (kaolinite).
   d. *Striations.* On crystal faces of quartz and pyrite, grooves or lines (termed striations) are commonly seen (Fig. 1-3). These striations reflect nonuniform crystal growth rates. Much finer more closely-spaced striations are typically found on cleavage planes of plagioclase. These types of striations are a result of inversions in the plagioclase crystal structure during growth.
   e. *Double Refraction.* Light which is transmitted through transparent calcite is split into two rays. As a result, objects

**Figure 1-3.** Photograph of intergrown pyrite crystals. Note coarse striations on crystal faces.

viewed through the mineral show a double image.

f. *Solubility in Hydrochloric Acid.* When a small amount of cold, dilute hydrochlo-

*CAUTION ABOUT TASTING SAMPLES*

ric acid (HCl) is placed on certain carbonate minerals, a chemical reaction occurs which dissolves the mineral. The reaction may be observed by a "bubbling" or effervescent "fizzing" in the acid drop (a result of the liberation of carbon dioxide, $CO_2$, by the reaction). Calcite will effervesce freely when the acid is placed on it while dolomite must be powered (thereby creating more surface area for reaction) before the effervesence is produced.

A specific mineral species may not display all of the physical properties listed above. In addition, due to small compositional variations within a mineral species, individual specimens may show slightly different aspects of each physical property. Therefore, it is good practice to examine two or three specimens to most completely describe a specific mineral's physical properties.

## EXERCISE 1

The following exercise will introduce the study of minerals. It will give you the basic experience in observing physical properties which is required before actual mineral identification can be undertaken.

1. Examine the physical property display sets available in your laboratory. Familiarize yourself with all of the physical properties which are described in the previous section.

2. Below is a list of 10 different materials. Indicate which are minerals and which are not minerals. Briefly explain the criteria for your identification.

| MATERIAL | MINERAL OR NONMINERAL | EXPLANATION |
|---|---|---|
| oyster shell | _____ | _____ |
| ice | _____ | _____ |
| synthetic diamond | _____ | _____ |
| sugar | _____ | _____ |
| window glass | _____ | _____ |
| wood | _____ | _____ |
| emerald | _____ | _____ |
| pearl | _____ | _____ |
| crude oil | _____ | _____ |
| iron ore | _____ | _____ |

3. Examine the following samples and indicate which are crystal aggregates, which are individual crystals and which are fragments of larger crystals.

| Sample | Crystalline Aggregate | Crystal | Crystal Fragment |
|---|---|---|---|
| _____ | _____ | _____ | _____ |
| _____ | _____ | _____ | _____ |
| _____ | _____ | _____ | _____ |
| _____ | _____ | _____ | _____ |
| _____ | _____ | _____ | _____ |
| _____ | _____ | _____ | _____ |
| _____ | _____ | _____ | _____ |
| _____ | _____ | _____ | _____ |

4. Describe the luster of the following samples. For those which have a nonmetallic luster, also use a descriptive adjective (as shown in Table 1-1) to further describe the character of the luster.

**Sample**                          **Luster**

_____     _____

_____     _____

_____     _____

_____     _____

5. For each of the specimens listed below, determine whether it does or does not have cleavage. If it does, list the number of cleavage directions and estimate the angles between the different sets of cleavage planes.

| Sample | Cleavage (yes or no) | Number of Directions | Angles Between Cleavage Planes |
|--------|----------------------|----------------------|--------------------------------|
| _____ | _____ | _____ | _____ |
| _____ | _____ | _____ | _____ |
| _____ | _____ | _____ | _____ |
| _____ | _____ | _____ | _____ |
| _____ | _____ | _____ | _____ |

6. Listed below are 5 samples. Determine their hardness and list in order of increasing hardness.

| Sample | Hardness | Listed in Increasing Hardness |
|--------|----------|-------------------------------|
| _____ | _____ | _____ |
| _____ | _____ | _____ |
| _____ | _____ | _____ |
| _____ | _____ | _____ |
| _____ | _____ | _____ |

# Exercise 2

# Mineral Identification

More than 2,000 different mineral species have been identified, however only 20 to 30 are abundant constituents of rocks. The purpose of this laboratory is to acquaint you with these common rock-forming minerals. The most diagnostic physical properties of the minerals which you will be examining are listed in the Mineral Identification Table at the end of this chapter.

The process of mineral identification is simplified if a systematic approach is used. Note that the Mineral Identification Table is divided into several sections, each listing mineral species which have the following physical properties in common:

1. Luster—metallic and nonmetallic.
2. Minerals with a nonmetallic luster are subdivided on the basis of light and dark color.
3. Minerals with a metallic luster are subdivided on the basis of their streak color.
4. The luster groups are broken down into smaller categories by comparison of their hardness to that of glass (less than glass, greater than glass, about the same as glass).
5. Individual mineral species are then listed according to the presence or absence of cleavage.

To identify an unknown mineral specimen, systematically describe its physical properties in the order listed above. The following example illustrates this technique. You observe that an unknown mineral specimen has a nonmetallic luster. This indicates that the unknown cannot belong to the class of minerals with a metallic luster. You observe that the specimen has a black color. This excludes the group of light-colored minerals with a nonmetallic luster. Next you check the unknown's hardness and find that it scratches glass. This narrows the possible choices to augite, hornblende, garnet or olivine (see Mineral Identification Table). The common color of olivine (light green) and garnet (dark brown to red) is not similar to the black color of the unknown and these two choices can be tentatively ruled out. Next you observe that the specimen has two directions of cleavage which are not at right angles. Inspection of the Mineral Identification Table indicates that neither olivine or garnet display cleavage, further substantiating your tentative conclusion made on the basis of color. The Table indicates that both augite and hornblende have two directions of cleavage, but that only in hornblende are the two cleavage directions not at right angles. Thus, the unknown mineral has been identified as hornblende. Following this example, any of the common rock-forming minerals may be easily identified.

## EXERCISE 2

In this exercise you are given samples of the more abundant rock-forming minerals. Your job is to completely describe and identify these minerals. Record your observations and identifications in the forms provided at the end of the chapter (be neat). Do not attempt to memorize all of the properties of each mineral, but, instead, learn to use the major physical properties for identification. It is important to describe each specimen in a systematic way using the following outline:

1. Determine the luster of the sample (metallic or nonmetallic).
2. If the sample has a metallic luster, what is its streak color?
3. If the sample has a nonmetallic luster, is it light or dark colored? What is its streak color?
4. What is the hardness of the sample?
5. Does the mineral have cleavage? If so, how many different sets of cleavage planes are present and what are the angles between the planes? What is the perfection of development of the cleavage planes?
6. What is the relative "heft" of the mineral (its specific gravity)?
7. Note any special physical properties which may be useful in identification (magnetism, taste, odor, diaphaneity, reaction with acid, etc.).

Compare your systematic list of physical properties with the Mineral Identification Table and select the mineral species whose physical properties most closely fits your description. Be certain to check several other specimens of the mineral in your laboratory to determine the extent of variations in its physical properties.

Mineral Identification Form

| Sample Number | Luster | Color | Streak | Hardness | Cleavage Number | Cleavage Angle | Special Properties | Mineral Name |
|---|---|---|---|---|---|---|---|---|
| 1 | | | | | | | | |
| 2 | | | | | | | | |
| 3 | | | | | | | | |
| | | | | | | | | |
| | | | | | | | | |
| | | | | | | | | |
| | | | | | | | | |
| | | | | | | | | |
| | | | | | | | | |
| | | | | | | | | |
| | | | | | | | | |
| | | | | | | | | |

Mineral Identification Form

| Sample Number | Luster | Color | Streak | Hardness | Cleavage | | Special Properties | Mineral Name |
|---|---|---|---|---|---|---|---|---|
| | | | | | Number | Angle | | |
| | | | | | | | | |
| | | | | | | | | |
| | | | | | | | | |
| | | | | | | | | |
| | | | | | | | | |
| | | | | | | | | |
| | | | | | | | | |
| | | | | | | | | |
| | | | | | | | | |
| | | | | | | | | |
| | | | | | | | | |
| | | | | | | | | |

12

**Mineral Identification Table**
Minerals With a Metallic Luster

| | Cl | Streak | Properties | Comments | Name and Composition |
|---|---|---|---|---|---|
| **H > Glass** | no | greenish black | Color *brass-yellow*; H=6-6.5; G=5; opaque. Commonly in masses of intergrown cubic crystals with *striated faces*. | Commonly known as "fool's gold." | **Pyrite** $FeS_2$ |
| | no | black | Color black; H=6; G=5.2; opaque; *strongly magnetic*. Usually occurs as massive aggregates. | Important ore of iron. | **Magnetite** $Fe_3O_4$ |
| **H ≈ Glass** | no | *red to red brown* | Color red-brown to black; H=5.5-6.5; G=5.3; opaque. May occur in aggregates of small, shiny flakes (specular variety) or as more massive, dull to earthy looking masses with a nonmetallic luster. | Important ore of iron. | **Hematite** $Fe_2O_3$ |
| | no | *yellow to yellow brown* | Color yellowish brown to black; H=1-5.5; G=3.5-4; opaque. May be found as porous aggregates of various hydrated iron oxides which appear dull or earthy with a nonmetallic luster. | Resembles rust, a common weathering product of iron-bearing minerals. | **Limonite** $FeO(OH) \cdot nH_2O$ |
| **H < Glass** | yes | gray to black | Color *silvery gray*; H=2.5; G=7.5; opaque. *3 perfect cleavages at 90°*. | Important ore of lead. | **Galena** PbS |
| | yes | black | Color silvery gray to black; H=1; G=2.3; opaque. *1 perfect cleavage. Has a greasy feel (marks fingers)*. | Used as a lubricant and as a source of carbon. | **Graphite** C |

H=hardness; Cl=cleavage (yes-d, has cleavage but may be difficult to see); G=specific gravity. The most distinctive physical properties of each mineral species are in italics.

13

## Minerals With a Nonmetallic Luster
## Generally Dark-Colored

| Cl | Properties | Comments | Name and Composition |
|---|---|---|---|
| **H > Glass** | | | |
| yes-d | Color *dark green to black*; H=5-6; G=3.2-3.4; translucent on edges; vitreous to resinous; *2 cleavages at 90°* (not always obvious). | Most common pyroxene, found in many rock types. | **Augite** (Pyroxene Group) Fe-Ca-Mg aluminosilicate |
| yes | Color *dark green to black*; H=5-6; G=3.0-3.4; translucent on edges; vitreous to slightly resinous; *2 good cleavages at 60° and 120°*. Samples often have a splintery appearance due to the intersecting cleavage planes. | Most common amphibole, found in many rock types. | **Hornblende** (Amphibole Group) Fe-Ca-Mg aluminosilicate |
| no | Color variable, often *red-brown*; *H=7*; G=3.5-4.3; translucent; vitreous to slightly resinous; no cleavage but may have subparallel fracture planes showing conchoidal patterns. *Often occurs as round crystals with diamond-shaped faces.* | Usually found in metamorphic rocks. Commonly used as a gem stone and also as an abrasive in sandpaper. | **Garnet** Fe-Ca-Mg-Mn aluminosilicate |
| no | Color *olive green*; H=6.5-7; G=3.3-4.4; translucent to transparent; vitreous. Commonly occurs in granular aggregates with a sugary texture. Individual grains may show conchoidal fracture. | Found mostly in mafic igneous rocks. Seldom found with quartz. | **Olivine** $(Mg, Fe)_2 (SiO_4)$ |
| **Glass ≈ H** | See Hematite and Limonite Under Metallic Luster | | |
| **H < Glass** | | | |
| yes | Color *light to dark brown*; H=2.5-3; G=2.8-3; thin sheets are transparent but smoky; vitreous; *1 perfect cleavage.* | Abundant in many rock types. | **Biotite** $K(Mg, Fe)_3 (AlSi_3)O_{10} (OH)_2$ |
| yes-d | Color *dark to light green*; H=2-2.5; G=2.6-2.9; translucent to transparent; vitreous to pearly; *1 perfect cleavage;* may have a pale green streak. Often occurs as fine-grained micaceous aggregates. | Most abundant in metamorphic rocks. | **Chlorite** $(Mg, Fe, Al)_6 (Al, Si)_4 O_{10} (OH)_8$ |

H=hardness; Cl=cleavage (yes-d, has cleavage but may be difficult to see); G=specific gravity. The most distinctive physical properties of each mineral species are in italics.

*(handwritten notes)*
SPHALERITE
YES Yellow Brown to Reddish Black
Vitreous Ye/bw White Streak
Good xtls

## Minerals With a Nonmetallic Luster
### Generally Light-Colored

| Cl | Properties | Comments | Name and Composition |
|---|---|---|---|
| **H > Glass** | | | |
| yes | Color variable, most common are *creamy white* to *flesh-red*; H=6; G=2.5; translucent; *2 good cleavages at nearly 90°*; vitreous to pearly. Often confused with plagioclase feldspar but lacks close-spaced striations. | Important member of the feldspar group, one of the most common rock-forming minerals. | **Orthoclase** (Potassium Feldspar) $KAlSi_3O_8$ |
| yes | Color creamy-white to gray; H=6; G=2.6-2.8; translucent; vitreous to pearly; *2 cleavages* (1 good and 1 poor) *at 90°*. Distinguishing *closely-spaced striations* on cleavage planes differentiate from orthoclase feldspar. | Important member of the feldspar group, one of the most common rock-forming minerals. | **Plagioclase** (Calcium-Sodium Feldspar) $NaAlSi_3O_8$-$CaAlSi_2O_8$ |
| no | Color variable but clear, smoky, purple and rose are most common; H=7; G=2.7; transparent to translucent; vitreous: Often found as well-formed, *six-sided crystals* with striations on some faces. Displays conchoidal fracture | Common rock-forming mineral, very resistant to weathering. | **Quartz** $SiO_2$ |
| no | Color variable, most common are brown, gray and creamy white; H=9; G=4; translucent; vitreous to adamantine. May occur as roughly six-sided forms with conchoidal fracture. Note hardness and specific gravity. | Commonly used as an abrasive; gem stones known as ruby or sapphire depending on color. | **Corundum** $Al_2O_3$ |
| **H ≈ Glass** | | | |
| yes | Color usually *bluish gray*; H=5-7; G=3.5; translucent; vitreous to pearly. Commonly formed as *bladelike crystals*. Hardness varies with respect to the direction scratched. *1 perfect cleavage.* | Found most often in metamorphic rocks. | **Kyanite** $Al_2SiO_5$ |
| **H < Glass** | | | |
| yes | Color varies widely, most common are clear, purple, yellow; H=4; G=3.2; transparent to translucent; *4 perfect directions of cleavage.* Due to perfection of cleavage, specimens often occur as tetrahedral shapes (these are not crystal forms however). May appear dark-colored. | High quality material finds use in lenses and prisms. | **Fluorite** $CaF_2$ |
| yes-d | Color clear, milky white or light yellow; H=3-3.5; G=4.5; transparent to translucent; vitreous to pearly; 2 directions of cleavage (1 perfect, 1 poor) at 90° (cleavage may be difficult to see). Note *unusually large specific gravity for a light-colored mineral.* | Major ore of barium. | **Barite** $BaSO_4$ |

H=hardness; Cl=cleavage (yes-d, has cleavage but may be difficult to see); G=specific gravity. The most distinctive physical properties of each mineral species are in italics.

Minerals With a Nonmetallic Luster
Generally Light-Colored (Ctd)

| Cl | Properties | Comments | Name and Composition |
|---|---|---|---|
| yes-d | Color is variable, often white, gray, pink, H=3.5-4; G=2.8; translucent; vitreous to pearly; *3 directions of cleavage, not at right angles* (may be difficult to recognize). Often found as crystalline aggregates. *Reacts with cold, dilute HCl very slowly, more rapid reaction if powdered.* | Major mineral component of some sedimentary rocks, used in cement and in steel manufacturing. | Dolomite (Ca, Mg)CO$_3$ |
| yes | Color variable, most commonly white to colorless; H=3; G=2.7; transparent to translucent; vitreous to oily; *3 directions of perfect cleavage not at right angles* (specimens will break into rhombohedrons as a result of this cleavage). *Reacts vigorously with cold, dilute HCl.* Transparent samples show strong *double refraction.* | Common in many rock types, but is most typical of sedimentary rocks, its principal use is in cement manufacturing and for making lime for fertilizers. | Calcite CaCO$_3$ |
| yes | Color light yellow to brown; H=2-2.5; G=2.5-3; thin sheets are transparent; vitreous, pearly, silky; elastic in thin sheets; *1 perfect direction of cleavage.* | Mostly found in metamorphic rocks, has been used for electrical insulators and in windows. | Muscovite KAl$_2$(AlSi$_3$)O$_{10}$(OH)$_2$ |
| yes | Color generally clear to milky white; H=2.5; G=2.2; transparent to translucent; vitreous, greasy, resinous; 3 directions of perfect cleavage at 90°. *Has a distinctive salty taste.* | Common table salt. | Halite NaCl |

H < Glass

H=hardness; Cl=cleavage (yes-d, has cleavage but may be difficult to see); G=specific gravity. The most distinctive physical properties of each mineral species are in italics.

Minerals With a Nonmetallic Luster
Generally Light-Colored (Ctd)

| Cl | Properties | Comments | Name and Composition |
|---|---|---|---|
| yes-d | Color variable, often creamy white, gray and yellow; $H=2$; $G=2.3$; transparent to translucent; vitreous to pearly or dull; 3 directions of cleavage (perfect in one direction and poor in the other) at 90° (may be difficult to see). Occurs as well-formed bladed crystals (selenite), or as dull, massive aggregates (alabaster). | Found mostly in sedimentary rocks, used for making plaster. | **Gypsum** $CaSO_4 \cdot nH_2O$ |
| yes-d | Color light green to silvery white; $H=1$; $G=2.7$; translucent; pearly to greasy; 1 perfect cleavage (may be difficult to see); *has a greasy feel.* Often occurs fine-grained, micaceous aggregates. | Powdered for use in toilet preparations. | **Talc** $Mg_3Si_4O_{10}(OH)_2$ |
| yes-d | Color white (often colored by impurities); $H=2$-$2.5$; $G=2.6$; opaque; *dull or earthy;* 1 perfect cleavage (may be difficult to see). Usually occurs in claylike aggregates which have an *earthy odor* when moist. | Principal mineral component in clays. Used extensively in making brick and porcelain products. | **Kaolinite** $Al_4Si_4O_{10}(OH)_8$ |
| no | Color white or yellowish brown; $H=1$-$3$; $G=2$-$2.5$; opaque; dull or earthy. Not really a mineral but a mixture of hydrous aluminum oxides of variable composition. Commonly found as *massive, earthy, claylike aggregates* or as small, *pea-shaped concretionary grains.* Hardness may be affected by secondary silica cement. | Major ore of aluminum, common weathering product in subtropical and tropical areas. | **Bauxite** $Al_2O_3 \cdot H_2O$ |

H < Glass

H=hardness; Cl=cleavage (yes-d, has cleavage but may be difficult to see); G=specific gravity. The most distinctive physical properties of each mineral species are in italics.

# Introduction to the Study of Rocks

A piece of the earth's crust is more likely to be a rock than a mineral. Although the majority of rocks are composed of minerals, some are formed totally of organic material (coal or limestone, for example), while others are formed of noncrystalline, inorganic material (such as obsidian). In order to discuss and study rocks we must have a common terminology for reference. Since no two rock specimens are exactly identical, it is necessary to establish some distinguishing criteria which can relate rocks which have generally similar characteristics.

One such characteristic is environment of formation and rocks have long been divided into three major categories on the basis of their modes of origin. *Igneous* rocks (from the Latin word *ignis,* fire) are those which have crystallized from a high-temperature melt, called a magma. This category includes all rocks formed from volcanic activity and also those which have formed from magmas which have cooled below the earth's surface. Another group of rocks are termed *sedimentary* (from the Latin word *sedimentum,* settling) and are formed from material which has accumulated at the earth's surface (generally deposited in water). The other major class of rocks is termed metamorphic (from the Greek word *metamorphosis,* to transform). This group includes rocks which have been texturally and mineralogically changed as a result of being exposed to temperatures and pressures which were different from that of their initial formation. Metamorphic rocks can form from igneous, sedimentary or other metamorphic rocks.

These three categories set apart rocks which form under widely different conditions. However, the distinction between igneous, sedimentary and metamorphic origins is not always obvious. For example, during some volcanic eruptions, large quantities of very fine volcanic debris (ash or dust) is blown into the atmosphere. Some of this ash often settles into lake bottoms and is compacted into a rock. Is the resultant rock igneous (it was formed of primary igneous material) or sedimen-

tary (it accumulated in a sedimentary fashion and was converted into rock by sedimentary processes)? Other such examples could be cited, and the ambiguity which they exemplify is a result of the somewhat arbitrary classification scale which geologists have established. However, in spite of the ambiguity, a genetic type of overall rock classification is superior to a descriptive one, where, for example, all green rocks would be considered as similar.

The environment of formation of a rock strongly influences the type and morphology (shape) of its mineral constituents. Therefore, the genetic classification of rocks into igneous, sedimentary and metamorphic categories implies certain common descriptive characteristics.

Most igneous rocks form at high temperatures by crystallization of minerals from a liquid magma. This mutual crystallization of minerals produces a complex intergrowth of crystals and yields a diagnostic igneous texture (Fig. 3-1). Because of the complex intergrowth of minerals, most igneous rocks are hard and dense. In addition, as they form at elevated temperatures, igneous rocks are composed only of minerals which are stable at high

**Figure 3-1.** Photograph of an igneous rock showing typical interlocking crystal texture.

temperatures (predominantly feldspars, quartz, hornblende, pyroxene, micas, and olivine).

Sedimentary rocks form at the earth's surface at generally low temperatures. Some sedimentary rocks (termed *clastic*) form from fragments of older rocks and minerals which have been eroded, transported (either by ice, wind or water) and then deposited. As a result of abrasion during transportation, the constituent fragments are generally rounded and, in contrast to igneous rocks, do not have an interlocking crystal texture (Fig. 3-2). Because they form from material which accumulates at the earth's surface, the chief mineral components of clastic sedimentary rocks are generally those which are most stable at low temperatures (primarily quartz and clays). Some relic high-temperature mineral fragments (such as orthoclase) may survive the transportation and deposition processes and be found in clastic sedimentary rocks.

Metamorphic rocks are produced from pre-existing rocks by changing the mineral composition and/or texture. The extent of textural and mineralogical modification is variable. Most of these changes take place at depth within the earth's crust and this reconstitution is generally accompanied by deformational movements. As a result, most metamorphic rocks have a layered (or *foliated*) appearance (Fig. 3-4). Most metamorphic rocks have a large percentage of micas (chlorite, biotite, and/or muscovite) which serves to define this layered texture. Metamorphic rocks generally form at temperatures intermediate between those of igneous and sedimentary environments and also have distinctive mineral constituents. Quartz, biotite, chlorite and muscovite are the most common metamorphic minerals. Feldspar is generally less abundant than in igneous rocks. Also, some mineral species (such as garnet, staurolite, kyanite and sillimanite) are almost entirely restricted to metamorphic rocks.

**Figure 3-2.** Photograph of a clastic sedimentary rock. Note rounded mineral and rock fragments and the lack of an interlocking crystal texture.

**Figure 3-3.** Photograph of a nonclastic sedimentary rock. Abundant fossil fragments clearly indicate the origin of this rock.

Another group of sedimentary rocks (termed *nonclastic*) are formed principally of the calcium carbonate (calcite) skeletal fragments of organisms (such as coral or shells), or of minerals which have been chemically precipitated from water by inorganic processes (mostly dolomite, halite, chert or gypsum). The former group is readily distinguished by the presence of skeletal fragments (Fig. 3-3). The latter group may have an interlocking mineral framework, but are commonly very fine-grained. In addition, they are generally composed of only one mineral species. The low-temperature nature of the mineral constituents serves to distinguish this group from igneous rocks.

**Figure 3-4.** Photograph of a foliated metamorphic rock.

## EXERCISE 3

This exercise will introduce the study and classification of rocks. Examine the rock specimens listed by your instructor and identify the principal mineral species in each (use the form provided at the end of this chapter). Remember that most mineral components of a rock have grown in competition for space and, as a result, are generally small and lack well-defined crystal shapes. However, the same physical properties you used previously to identify larger mineral specimens can be used to identify the smaller grains. True, the task of identification may be difficult because of the small size of the minerals, but with patience, correct identification is possible.

Once you have described the mineralogy of the samples, segregate them into igneous, sedimentary (clastic or nonclastic) and metamorphic groups. List the most distinguishing characteristics which you used for this classification. Remember the following important properties of the three major rock groups.

I.  Igneous Rocks
    A.  Grain size and color are variable.
    B.  Usually hard and dense.
    C.  Characteristically have an interlocking crystal texture.
    D.  Are made up of minerals which are stable at high temperatures (mostly quartz, feldspars, hornblende, pyroxene and biotite).

II.  Sedimentary Rocks
    A.  Clastic Group.
        1.  Grain size and color are variable.
        2.  Made up of variably rounded fragments without an interlocking crystal texture.
        3.  Principal mineral constituents are quartz and clays.
    B.  Nonclastic Group.
        1.  Grain size and color are variable.
        2.  May have abundant skeletal fragments or a massive, microcrystalline texture.
        3.  Principal mineral constituents are calcite, dolomite, halite, chert and gypsum.

III.  Metamorphic Rocks.
    A.  Grain size and color are variable.
    B.  Most are layered (foliated).
    C.  Most abundant minerals include biotite, muscovite, chlorite, and quartz. Feldspars are less common than in igneous rocks.
    D.  May have diagnostic metamorphic minerals such as garnet, staurolite, kyanite and sillimanite.

## Rock Identification Form

| Sample Number | Principal Mineral Components | Rock Type (Igneous, Sedimentary, Metamorphic) | Distinguishing Characteristics |
|---|---|---|---|
| | | | |
| | | | |
| | | | |
| | | | |
| | | | |
| | | | |

| Rock Identification Form | | | |
|---|---|---|---|
| Sample Number | Principal Mineral Components | Rock Type (Igneous, Sedimentary, Metamorphic) | Distinguishing Characteristics |
| | | | |
| | | | |
| | | | |
| | | | |
| | | | |
| | | | |

22

# Exercise 4

# Igneous Rocks

Igneous rocks form when hot, molten material (*magma*) cools and solidifies. As cooling takes place, individual minerals begin to grow (*crystallize*) and form an interlocking crystal mosaic. The physical properties of a specific igneous rock depends primarily upon two factors: (1) the chemical composition of the magma from which it crystallized; and (2) the physical environment in which it crystallized. For example, a sodium-rich igneous rock cannot form by the cooling and crystallization of a magma which contained little sodium. In addition, a magma which cooled and crystallized rapidly at the earth's surface should appear distinctly different from one which slowly solidified at depth within the earth's crust.

Original magma composition is reflected in the type and relative amounts of mineral species in an igneous rock. Magmas which are rich in aluminum, silicon, sodium and potassium are generally termed *felsic*. These magmas crystallize to form generally light-colored igneous rocks composed primarily of quartz, orthoclase and sodium-plagioclase with minor amounts of biotite, muscovite and/or hornblende. On the other hand, magmas which contain abundant iron, magnesium and calcium (the *ferromagnesian* elements) are termed *mafic*. These crystallize to form generally dark-colored igneous rocks made up mostly of olivine, pyroxene and calcium-plagioclase along with minor amounts of hornblende. Magmas which are unusually rich in the ferromagnesian elements are termed *ultramafic* and generally crystallize to form igneous rocks made up largely of olivine, pyroxene and/or calcium-plagioclase. There is a wide variation in initial magma composition and many fall between the mafic and felsic end-members. Crystallization of magmas of intermediate composition yields igneous rocks which contain sodium+calcium—plagioclase, hornblende and biotite along with minor quartz and/or pyroxene. The color of these intermediate rocks is generally between the light-colored felsic and dark-colored mafic igneous rocks.

On the basis of chemical composition (reflected in mineral constituents), igneous rocks may be divided into four major groups:

1. Light-colored felsic varieties.
2. Intermediate varieties.
3. Dark-colored mafic varieties.
4. Dark-colored ultramafic varieties.

Magmas which solidify deep within the earth's crust are termed *intrusive*. They generally cool very slowly which allows for the growth of large crystals. Magmas which reach the earth's surface are termed *extrusive*. These cool rapidly and generally do not crystallize abundant large crystals. Thus, the size of constituent mineral grains is an important clue to the cooling history and, therefore, the environment of formation of an igneous rock. Because of this significance, igneous rock textures are generally separated into two major categories on the basis of grain size: (1) coarse-grained, termed *phaneritic* (mineral constituents generally large enough to be seen without a microscope); and (2) fine-grained, called *aphanitic* (mineral constituents too small to be seen without the aid of a microscope). These two textural types are illustrated in Figure 4-1.

If a magma has had a complex cooling history involving an early period of slow cooling followed by a later phase of more rapid cooling, it is likely that the resultant igneous rock would contain mineral grains of markedly different sizes. Minerals which formed during the early period of slow cooling would tend to be larger than those which grew during the later period of more rapid cooling. This type of texture is rather common in igneous rocks and is termed *porphyritic*. The larger mineral grains are termed *phenocrysts*. The smaller grains are termed the *matrix* or *groundmass*. If the matrix grains are aphanitic, the texture is termed *aphanitic porphyry* (Fig. 4-2A). If the matrix grains are phaneritic, the texture is termed *phaneritic prophyry* (Fig. 4-2B).

A

B

**Figure 4-1.** Photographs of igneous rocks illustrating texture: (A) equidimensional phaneritic (individual mineral grains may be identified); (B) equidimensional aphanitic (individual grains may not be identified).

A

B

**Figure 4-2.** Photographs of igneous rocks illustrating texture: (A) aphanitic porphyry (large grains in an aphanitic matrix); (B) phaneritic porphyry (large grains in a phaneritic matrix).

Under special conditions of cooling, several other igneous textures may develop. If magmas cool very rapidly at the earth's surface, there may not be time for the growth of any mineral species. Although solid, the resultant rock would actually be a supercooled, noncrystalline liquid (similar to glass). The texture of these igneous rocks (such as obsidian) is glassy (Fig. 4-3A). Magmas often contain appreciable amounts of gas (largely water) which forms bubbles (or *vesicles*) in the magma. Because of the rapid cooling of extrusive magmas, these bubbles are often trapped during crystalliza-

tion, leaving the igneous rocks with a spongy or *vesicular* texture (Fig. 4-3B). These types of rocks are generally named *pumice* if light-colored or *scoria* if dark-colored. Fine-grained material (ash or dust) which is blown out of a volcano may settle from the atmosphere to form rocks composed of fragmental igneous grains. These typically have angular or broken textures and are termed *tuffs* or *breccias*.

To classify an igneous rock both mineralogical (chemical) and textural properties are used (illustrated in the Igneous Rock Classification Table).

**Figure 4-3.** Photographs of igneous rocks illustrating texture: (A) glassy (obsidian—note conchoidal fracture); (B) vesicular (scoria).

Most felsic magmas crystallize to form either coarse-grained (phaneritic) rocks termed *granite* or fine-grained (aphanitic) rocks termed *rhyolite*. Depending upon the complexity of a magma's cooling history, either rock type may have a porphyritic texture. On the other hand, magmas with a mafic composition crystallize to form coarse-grained rocks named *gabbro* or fine-grained rocks called *basalt* (either of which may be porphyritic). Magmas with an intermediate composition crystallize to form coarse-grained igneous rocks named *diorite* or fine-grained rocks named *andesite*. Again, either may be porphyritic. Magmas which contain an unusually large percentage of ferromagnesium elements (the ultramafic group) are most often represented by coarse-grained igneous rocks named *dunite* (if mostly olivine), *pyroxenite* (if mostly pyroxene) or *anorthosite* (if mostly calcium-plagioclase). Due to their extremely high-temperature of formation, these ultramafic rocks commonly do not have any fine-grained equivalents which form at the earth's surface.

Igneous rock types are texturally and mineralogically (chemically) gradational, and the rock names on the Igneous Rock Classification Table are generalized.

# CLASSIFICATION OF IGNEOUS ROCKS

**General Color:** *Felsic* — Light | Intermediate | *MAFIC* — Dark

**Mineralogy — % Volume Abundance of Minerals Present:** 100%, 75%, 50%, 25%

Minerals: Orthoclase · Quartz · Plagioclase (Ca, Na) · Olivine · Pyroxene · Hornblende · Micas · Mostly Olivine, Pyroxene or Calcium-Plagioclase

| TEXTURE | Light (Felsic) | Intermediate | Dark (Mafic) | Dark (ultramafic) |
|---|---|---|---|---|
| Phaneritic | Granite | Diorite | Gabbro | Dunite / Pyroxenite / Anorthosite |
| Phaneritic Porphyry | Porphyritic Granite | Porphyritic Diorite | Porphyritic Gabbro | |
| Aphanitic | Rhyolite | Andesite | Basalt | |
| Aphanitic Porphyry | Porphyritic Rhyolite | Porphyritic Andesite | Porphyritic Basalt | |
| Glassy | Obsidian | Obsidian | | |
| Vesicular | Pumice | | Scoria | |
| Pyroclastic | Tuff, Breccia | | | |

Handwritten annotations: *SYENITE*; *FELSITIC APHANITIC NO PHENOCRYSTS*

# EXERCISE 4

The purpose of this laboratory is to acquaint you with the more common types of igneous rocks found in the earth's crust. Carefully examine and describe each rock sample listed by your instructor. Use the accompanying forms (be neat). Classify each sample using the Igneous Rock Classification Table.

To identify an igneous rock, the following general outline should be followed:

1. Determine the rock's texture.

2. Describe the color of the sample and determine if it may be felsic, intermediate, mafic or ultramafic in composition.

3. a. If the rock has a phaneritic or a porphyritic phaneritic texture, identify the minerals present and estimate their approximate volume percentages (i.e., how much of the rock is quartz, how much is orthoclase feldspar, etc.). As a guide for identification of minerals in igneous rocks, Table 4-1 lists the eight most common igneous minerals and their distinguishing physical properties. Particular emphasis should be placed on a determination of the following:

   (1) Amount of quartz.
       (a) no quartz = gabbro or ultramafic.
       (b) less than 10% quartz = diorite.
       (c) 10%-40% quartz = granite.

   (2) Amount and type of feldspar.
       (a) 25%-75%, mostly orthoclase = granite.
       (b) 25%-50%, mostly plagioclase = diorite.
       (c) 15%-50%, only plagioclase = gabbro.
       (d) all plagioclase = ultramafic (anorthosite).

   (3) Amount and type of ferromagnesian minerals (olivine, pyroxene, hornblende, biotite, muscovite).
       (a) less than 10%, mostly biotite and/or muscovite, minor hornblende = granite.
       (b) 10%-40%, mostly hornblende, minor biotite and/or muscovite and/or pyroxene = diorite.
       (c) 40%-80%, mostly pyroxene and/or olivine, minor hornblende = gabbro.
       (d) greater than 80% = ultramafic.
           1. mostly olivine = dunite.
           2. mostly pyroxene = pyroxenite.

   b. If the rock is aphanitic, the minerals will be too small for identification and a generalization from rock color must be used as a guide to the overall chemical composition.

   c. If the rock is an aphanitic porphyry and has visible phenocrysts, these should be identified. The mineral species of the phenocrysts may offer a diagnostic clue to the chemical nature of the parent magma.
      (1) If the phenocrysts are olivine or pyroxene, the rock is a basalt.
      (2) If the phenocrysts are hornblende, the rock is an andesite.
      (3) If the phenocrysts are quartz or orthoclase, the rock is rhyolite.

4. Combine the textural and mineralogical characteristics and compare to the Igneous Rock Classification Table.

The following will serve as an example: you observe that a rock sample has large, pink phenocrysts set in a light gray matrix in which individual grains may be recognized. Further examination reveals that the phenocrysts are orthoclase feldspar and total more than 30 volume percent of the rock. The matrix is composed of 35% quartz, 15% orthoclase feldspar, 15% plagioclase and 5% biotite. Examination of the Igneous Rock Classification Table indicates that the rock is a porphyritic granite.

**TABLE 4-1**

Characteristic Appearance of Minerals
Common in Igneous Rocks

| Mineral | Diagnostic Physical Properties |
|---------|-------------------------------|
| Orthoclase Feldspar | 1. Generally equant grains.<br>2. Usually white or pink.<br>3. Two directions of cleavage. |
| Plagioclase Feldspar | 1. Lath-shaped grains.<br>2. Usually white or gray, may be slightly greenish.<br>3. Two directions of cleavage.<br>4. Striations on cleavages. |
| Quartz | 1. Usually interstital, poorly defined grains.<br>2. Colorless, glassy appearance.<br>3. No cleavage. |
| Olivine | 1. Small equidimensional grains.<br>2. Pale green, glassy.<br>3. No cleavage. |
| Hornblende (amphibole) | 1. Lath-shaped grains.<br>2. Black or dark greenish-black.<br>3. Two directions of cleavage not at right angles. |
| Augite (pyroxene) | 1. Equidimensional, blocky grains (may also be lath-shaped).<br>2. Black or dark greenish-black.<br>3. Two directions of cleavage at right angles. |
| Biotite | 1. Flat, flaky grains.<br>2. Black or dark brown.<br>3. 1 perfect cleavage. |
| Muscovite | 1. Flat, flaky grains.<br>2. Light green to silvery white.<br>3. 1 perfect cleavage. |

| Igneous Rock Identification Form | Sample Number | Texture | Color | Minerals and Percent Abundance | Rock Name |
|---|---|---|---|---|---|
| | | | | | |
| | | | | | |
| | | | | | |
| | | | | | |
| | | | | | |
| | | | | | |

# Igneous Rock Identification Form

| Sample Number | Texture | Color | Minerals and Percent Abundance | Rock Name |
|---|---|---|---|---|
| | | | | |
| | | | | |
| | | | | |
| | | | | |
| | | | | |
| | | | | |

# Exercise 5

# Sedimentary Rocks

The textural and mineralogical properties of any rock are strongly influenced by the environment in which it formed. Igneous rocks form by the solidification of magmas and contain minerals which are stable under high-temperature conditions. When igneous rocks are exposed at the earth's surface these high-temperature minerals are no longer stable. As a result, physical and chemical changes occur which lead to the formation of new mineral species which are stable in the surface environment. These mineralogical and textural changes are termed *weathering*. The extent of weathering of a rock is controlled by environmental conditions (such as temperature and availability of water) and its mineralogy. Ferromagnesian minerals (olivine, pyroxene, hornblende, and biotite) form at high temperatures and commonly weather rapidly (their susceptibility to decomposition is given by the order in which the minerals are listed; olivine most susceptible, biotite least). The more felsic minerals (sodium-plagioclase feldspar, orthoclase feldspar, and quartz) form at somewhat lower temperatures and weather more slowly (their susceptibility is indicated by the order in which they are listed).

The minerals which are formed during the weathering process and the remaining partially decomposed rock and mineral fragments are generally transported away from the immediate site of weathering. During transportation, further chemical breakdown of minerals may occur. Eventually, the weathering products (usually clays) and those original minerals which are most stable under surface conditions (primarily quartz) are deposited and accumulate to form *clastic* sediments. Other types of sediment, termed *nonclastic,* form as a result of the organic or inorganic precipitation of elements which were actually dissolved during the weathering and transportation processes.

As sediment accumulates, the weight of overlying material generally compacts buried sediment layers. Recrystallization of certain minerals (such as clays and calcite) may occur. In addition, secondary low temperature minerals (such as quartz, calcite and various types of iron oxides) may be deposited in void spaces between individual clastic grains. Growth of these secondary minerals serves to cement sedimentary grains into a rigid framework. As a result of this compaction, recrystallization and cementation, sediment is converted (or *lithified*) into sedimentary rocks.

Sedimentary rocks are difficult to classify on a chemical basis because the debris supplied from pre-existing rocks is quite varied. However, on a textural basis, sedimentary rocks may be divided into two basic groups: (1) *clastic* rocks composed, in part, of fragments of pre-existing rocks and minerals; and (2) *nonclastic* rocks made up of minerals which formed during organic or inorganic precipitation of dissolved chemical species.

Clastic sedimentary rocks may be further classified on the basis of the average size of their constituent grains or fragments. A four-fold size division is sufficient for a general rock classification (Table 5-1; Fig. 5-1). If grains are generally larger than 2 mm (boulder-, cobble-, pebble-sized), the rock is called *coarse-grained.* If they are between 0.062 and 2 mm (sand-sized), they are termed *medium-grained.* If the particles are between 0.005 and 0.062 mm (silt-sized; individual grains too small to be seen—rock feels "gritty" when rubbed), the rock is termed *fine-grained.* If the constituent particles are less than 0.005 mm (clay-sized; individual grains too small to be seen—rock feels smooth when rubbed), the rock is termed very fine-grained. Each of these four size divisions may be subdivided on the basis of the morphology (shape) and composition (mineralogy) of the constituent grains.

Coarse-grained rocks are termed *conglomerates* if constituent grains are rounded and smooth (Fig. 5-2A). If the grains are angular, the rock is termed *breccia* (Fig. 5-2B). Medium-grained rocks are termed *sandstone* if most of the grains are

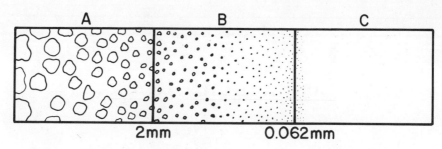

**Figure 5-1.** Size scale for use in classifying clastic sedimentary rocks: (A) coarse-grained; (B) medium-grained; (C) fine- and very fine-grained.

quartz, *arkose* if the grains are quartz and orthoclase feldspar, and *graywacke* if quartz, rock fragments and clays are the principal grain components. Fine-grained rocks are termed *siltstone* and very fine-grained rocks are called *shale* (both made up largely of clays and quartz but small grain size precludes identification). An outline of this classification is listed in the Clastic Sedimentary Rock Classification Table.

Nonclastic sedimentary rocks are formed by the precipitation of elements which are dissolved in water. They commonly have an interlocking crystal texture similar to that of igneous rocks. Grain size varies from *crystalline* (where individual crystals may be seen without a microscope) to *microcrystalline* (where the individual crystals are too small to be seen without the aid of a microscope). Regardless of grain size, the crystal framework is

made up mostly of one mineral species and the nonclastic sedimentary rocks are primarily classified on the basis of their composition (mineralogy).

Rocks composed of chalcedony (a variety of microcrystalline quartz, $SiO_2$) are named *chert*. *Rock salt* is a rock composed mainly of halite (NaCl). Rocks composed mostly of *dolomite* or *gypsum* are given the mineral names. Rocks composed largely of calcite ($CaCO_3$) are generally broken down into several categories on the basis of special textural characteristics. Massive crystalline calcite rocks are named *limestone*. If fossil fragments are abundant and firmly implanted in the calcite matrix the rock is named *fossiliferous limestone* (see Fig. 3-3). Finely laminated crystalline calcite rocks are called *travertine*. Coarse-grained rocks which are made up of loosely-cemented fossil

**TABLE 5-1**

Clastic Sedimentary Rock Classification

| Grain Size | Characteristics | Rock Name |
|---|---|---|
| Coarse-Grained (larger than 2mm) | Rounded rock or mineral fragments | **Conglomerate** |
| | Angular rock or mineral fragments | **Breccia** |
| Medium-Grained (0.062-2mm) | Composed mostly of quartz | **Sandstone** |
| | Composed mostly of quartz and orthoclase feldspar | **Arkose** |
| | Composed of clays, quartz and rock fragments | **Graywacke** |
| Fine-Grained (0.005-0.062mm) | Composed of quartz and clays (feels "gritty") | **Siltstone** |
| Very Fine-Grained (smaller than 0.005mm) | Composed of quartz and clays (feels smooth) | **Shale** |

fragments are named *coquina* (actually has a clastic texture). Microcrystalline calcite rocks are named *micrite.* If composed of soft, microcrystalline fossils, a rock is named *chalk.* A general outline for the classification of nonclastic sedimentary rocks is presented in Table 5-2.

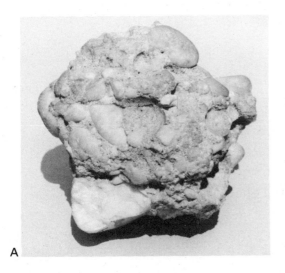

**Figure 5-2.** Typical appearance of coarse-grained clastic sedimentary rocks: (A) conglomerate (note rounded shape of constituent fragments); (B) breccia (note broken, angular shape of constituent fragments).

**TABLE 5-2**
Nonclastic Sedimentary Rock Classification

| Principal Mineral Component | General Characteristics | Rock Name |
|---|---|---|
| Halite | variably crystalline | Rock Salt |
| Gypsum | variably crystalline | Gypsum |
| Dolomite | variably crystalline to micro-crystalline | Dolomite |
| Calcite | massive, variably crystalline | Limestone |
| | massive, variably crystalline with fossil fragments | Fossiliferous Limestone |
| | finely banded, variably crystalline | Travertine |
| | abundant, poorly-cemented fossils | Coquina |
| | massive, microcrystalline | Micrite |
| | massive, soft, abundant micro-crystalline fossils | Chalk |

OOLITES SM. ROUND CONCENTRIC GRAINS LIKE SM. MARBLES          OOLITIC LS

33

# EXERCISE 5

This exercise will familiarize you with the more abundant types of sedimentary rocks. Carefully examine and describe each sample listed by your instructor. Use the accompanying forms (be neat). Identify each sample by comparing your descriptions with the Sedimentary Rock Classification Tables.

To identify a sedimentary rock, the following general procedure should be followed:

1. Describe the texture of the rock. Determine if it is clastic (composed of distinct mineral and/or rock fragments or grains) or nonclastic (has an interlocking crystal texture).

2. If the rock is nonclastic, identify the minerals present. Use the standard mineral identification techniques (hardness, reaction with acid, cleavage, etc.). If the principal mineral species is chalcedony, dolomite, gypsum or halite, the rock is identified. If composed primarily of calcite, further description is necessary. Note the following:
   a. Can individual crystals be observed (is the texture crystalline or microcrystalline)?
   b. Is the rock massive or laminated?
   c. Are fossil fragments present and, if so, are they well cemented?

3. If the rock has a clastic texture:
   a. Determine the average grain size.
      (1) If fine-grained (rock feels "gritty" when rubbed) it is termed a siltstone (individual grains too small for identification).
      (2) If very fine-grained (feels smooth when rubbed) it is a shale (individual grains too small for identification).
   b. If the rock has a medium grain size, identify the principal mineral species present.
   c. If the rock is coarse-grained, describe the general shape of constituent grains.

**Note.** Because of the common presence of secondary cementing minerals, caution must be used in identifying mineral species in clastic sedimentary rocks. Always make your identification tests with the aid of a binocular microscope and make certain you are identifying the constituent grains and not the cement.

## Sedimentary Rock Identification Form

| Sample Number | Texture | Mineralogy | Other Features | Name |
|---|---|---|---|---|
| | | | | |
| | | | | |
| | | | | |
| | | | | |
| | | | | |

## Sedimentary Rock Identification Form

| Sample Number | Texture | Mineralogy | Other Features | Name |
|---|---|---|---|---|
| | | | | |
| | | | | |
| | | | | |
| | | | | |
| | | | | |
| | | | | |

# Metamorphic Rocks

The earth is constantly in a dynamic state of change (although exceedingly slow by our standards of time) and rocks seldom remain in the environment where they formed for long periods of time. Igneous rocks which crystallized from a magma deep within the earth's crust may be exposed by the erosion of overlying rocks. Minerals of the igneous rock are unstable when exposed to the physical environment at the earth's surface and weathering occurs. Similarly, if sedimentary or extrusive igneous rocks are buried by a very thick accumulation of overlying material, they will experience temperature and pressure conditions different from those which characterized the surface environment in which they originally formed. Mineral species which were stable in the rocks become unstable, and new mineral species will crystallize. It is likely that the initial rock will become more compact because of the increased load pressure supplied by overlying material. This results in a rearrangement of constituent minerals. These mineralogical and textural changes caused by the increased temperature and pressure resulting from burial are collectively termed *regional metamorphism.*

The extent or intensity of recrystallization and reorientation during burial is termed the *grade* of metamorphism. It varies depending upon the depth of burial. Rocks which experience very deep burial may be exposed to temperatures and pressures which approximate that of the intrusive igneous environment (high metamorphic grade). On the other hand, lower-grade metamorphic changes (only slight recrystallization and reorientation) develop in rocks which are not so deeply buried.

Metamorphic rocks may also develop around igneous intrusions where heat and escaping igneous fluids produce textural and mineralogical changes in nearby rocks. This type of metamorphic alteration is more restricted in extent than that which results from deep burial and is termed *contact metamorphism.* Any rock may be affected by regional or contact metamorphism regardless of its original nature. However, intrusive igneous rocks and previously metamorphosed rocks (both of which form in high temperature and pressure environments) are generally less susceptible to metamorphic alteration than sedimentary or extrusive igneous rocks.

Metamorphic alteration of a rock includes three interrelated processes: (1) *crystallization* of new mineral species which are stable under the metamorphic temperature and pressure conditions; (2) *rotation* and *deformation* of minerals; and (3) original grains which remain stable under the metamorphic temperature and pressure conditions generally *recrystallize* into larger grains. The combined effects of these three processes is to produce a metamorphic rock which differs from the original rock (or *protolith*) by having greater crystallinity, being generally harder, and having a new texture which records evidence of deformation.

As with sedimentary rocks, classification of metamorphic rocks on the basis of chemical composition is difficult because of the extreme compositional variability of protolith rocks. Also, many minerals found in metamorphic rocks are common in other types of rocks (quartz, feldspar, hornblende, muscovite, and biotite). However, in general, biotite and muscovite are more abundant in metamorphic rocks and may be accompanied by certain mineral species which are typically found only in metamorphic rocks (including garnet, chlorite, sillimanite, kyanite, and staurolite). In addition, as metamorphic processes involve compaction together with crystallization and/or recrystallization, they often result in a pronounced preferred orientation of mineral grains. This three-dimensional orientation of constituent minerals is termed a *foliation* texture (Fig. 6-1). Platy minerals (such as chlorite, muscovite and biotite) are most susceptible to this three-dimensional orientation. The presence or absence of this texture (*foliated* or *nonfoliated* respectively) serves as a primary basis for the classification of metamorphic rocks.

Foliated metamorphic rocks have a strong parallel alignment of constituent minerals. They are subdivided on the basis of the nature of this orientation (Table 6-1; Fig. 6-1). Rocks which have experienced high-grade metamorphic conditions are usually coarse-grained and commonly have a well-defined parallel orientation of minerals. As a result of high temperatures, some minerals (quartz and feldspar) actually melt during metamorphism and segregate to form small, mineralogically distinct bands which parallel the preferred mineral orientation. Thus, most high-grade metamorphic rocks generally have a banded appearance (Fig. 6-1D) and are termed *gneisses.* Low grade metamorphic rocks are generally finer-grained (may be difficult to see individual mineral grains) although they generally show a marked preferred orientation of constituent platy minerals (chlorite and/or mus-covite). Because of the common abundance of platy minerals and their marked preferred orientation, low-grade metamorphic rocks are generally very *fissile* (break easily into thin plates). Very fine-grained fissile rocks are named *slate* (Fig. 6-1A) and fine-grained fissile rocks are named *phyllite.* Phyllites usually have surfaces with a general sheen or luster due to the reflection off musco-vite and/or chlorite which is aligned parallel to breakage surfaces (Fig. 6-1B). In addition, phyllite surfaces are commonly crumpled or contorted, whereas slate typically has more even breakage surfaces (Fig. 6-1B). Intermediate-grade metamorphic rocks generally have a medium- to coarse-grained texture and are well-foliated (Fig. 6-1C). These rocks are named *schist.* They are generally more severely crumpled and contorted than the low grade rocks and also contain abundant biotite.

A

B

C

D

**Figure 6-1.** Photographs of foliated metamorphic rocks: (A) slate (low-grade), very fine-grained, dull breakage surfaces, dark colored; (B) phyllite (moderate grade), fine-grained, contorted breakage surfaces with lusterous sheen; (C) schist (high-grade), medium-grained (most minerals may be identified), contorted breakage surfaces; (D) gneiss (very high-grade), coarse-grained (all minerals may be identified), light and dark bands nearly parallel foliation.

# EXERCISE 6

In this laboratory you will study the more common types of metamorphic rocks. Carefully examine and describe each sample listed by your instructor. Use the accompanying forms (be neat). Identify each rock using the Metamorphic Rock Classification Table.

To identify a metamorphic rock, the following general procedure should be followed:

1. Determine if the rock is foliated or nonfoliated (is there a preferred orientation of mineral grains).

2. If nonfoliated, identify the principle mineral species present.
   a. Quartz = quartzite.
   b. Calcite = marble.
   c. Hornblende and plagioclase = amphibolite (generally dark-colored, may have salt and pepper appearance).
   d. Fine-grained, no minerals can be seen = hornfels (massive, dark-colored).

3. If foliated, note the following:
   a. If individual mineral grains cannot be observed.
      (1) Generally dark-colored, fissile, dull surfaces = slate.
      (2) Lighter-colored, fissile, lustrous surfaces (may be contorted) = phyllite.
   b. If individual minerals can be observed, identify the principle mineral species.
      (1) Light and dark-colored bands, hornblende, quartz, biotite and feldspars may be present = gneiss (add name of principle mineral species present).
      (2) No well-developed banding, generally contorted foliation, biotite, muscovite, quartz, feldspar, garnet, kyanite, sillimanite, staurolite may be present = schist (add names of principle mineral species present).

## Metamorphic Rocks Identification Form

| Sample Number | Color | Foliated, Nonfoliated | Other Features | Mineralogy | Rock Name |
|---|---|---|---|---|---|
| | | | | | |
| | | | | | |
| | | | | | |
| | | | | | |
| | | | | | |
| | | | | | |

Metamorphic Rocks Identification Form

| Sample Number | Color | Foliated, Nonfoliated | Other Features | Mineralogy | Rock Name |
|---|---|---|---|---|---|
| | | | | | |
| | | | | | |
| | | | | | |
| | | | | | |
| | | | | | |

# Part 2

# GEOMORPHOLOGY
## A STUDY OF LANDFORMS

Agassiz, 1847

# Exercise 7

# Introduction to Topographic Maps

A map is a two-dimensional representation of a portion of the earth's surface. There are many different kinds of maps and most everyone has had some experience in map use (road maps or simple sketch maps showing directions to get to a particular point). All maps have one thing in common, they present a reduced image of a larger area. The amount of reduction (the *scale*) is variable and may be expressed in several ways:

1. Verbal Scale. The scale is expressed as a specific distance on the map equal to a specific distance on the ground. For example, one inch on the map equals one mile on the ground.

2. Graphic Scale. The scale consists of a calibrated bar or line which represents a specified distance on the ground. For example, a one-inch bar on the map is labeled as representing one mile on the ground. This is the usual type of scale on road maps.

3. Fractional Scale. The scale is expressed as a fixed ratio (termed the *representative fraction*) between a distance measured on the map and an equal distance measured on the earth's surface. For example, a fractional scale of 1:62,500 indicates that one distance unit on the map (whether inches, feet, yards, etc.) equals 62,500 of the *same* units on the surface of the earth.

In addition to scale, all maps must have some method of indicating location. The most common method is by reference to the *latitude–longitude* system which is the basic means of locating a point anywhere on the earth. Latitude is the angular distance north or south of the equator (0° latitude) and varies from 0-90° north and 0-90° south. Lines of latitude are frequently called *parallels* and encircle the earth from east to west. Longitude is the angular distance measured east or west of Greenwich, England (0° longitude) and ranges from 0-180° west and 0-180° east. Lines of longitude are termed *meridians* and encircle the earth from north

to south. The units of latitude and longitude are degrees, minutes (60 minutes = 1 degree) and seconds (60 seconds = 1 minute). Using this system, the precise location of any point on the earth may be given. For example, the coordinates which locate the University of Georgia geology building are lat 33°56′58″N–long 83°22′29″ W.

The maps which we will be using in the laboratory are called topographic maps. They show the three-dimensional shape of the earth's surface as well as horizontal distances. If the borders of the topographic map are 15′ of latitude and longitude, they are called 15′ quadrangles and are commonly published at a scale of 1:62,500. If the borders are 7.5′ of latitude and longitude, they are termed 7.5′ quadrangles and are commonly published at a scale of 1:24,000. Special symbols are used to designate various natural and cultural features on topographic maps. The enclosed description sheet and your laboratory instructor will help to explain these symbols.

Most topographic maps of the United States are divided into nine rectangles of equal size (Fig. 7-1). Each rectangle may be identified by its position with respect to the entire map (such as north-

| NW | NC | NE |
|----|----|----|
| WC | C | EC |
| SW | SC | SE |

**Figure 7-1.** Diagram illustrating the notation for location by rectangles on a topographic map: NE = northeast rectangle, WC = west central rectangle, C = central rectangle, etc.

west, east-central, etc.) and a convenient way of locating a point on a map is to designate the rectangle in which the point is located. Just as each rectangle of the map is identified, so may segments within each larger rectangle be identified. For example, the precise location of point A in Figure 7-1 would be given as NE-SC (northeast corner of south-central rectangle).

Topographic maps are fundamental to earth science as they show the three-dimensional configuration of the earth's surface by means of ==contours== (Figs. 7-2, 7-3, 7-4). Contour lines may be visualized as lines of intersection between a series of equally-spaced horizontal planes and the ground surface. In Figure 7-2, horizontal planes 10 feet apart are shown cutting through an asymmetric hill. The hill is represented on the accompanying topographic map by a series of elliptical contour lines. The line XY is termed the ==profile line== (or line of the section) from which the hill is viewed from the side. Figure 7-3 shows a view of a ==depression== and its representation by contour lines. Note that the ==contours== outline the same elliptical form but are "==ticked" (hachured)== to indicate that the landform is a depression and not a hill. Figure 7-4 illustrates how hachured contour lines are used in combination with normal contour lines to indicate both hills and depressions within an area.

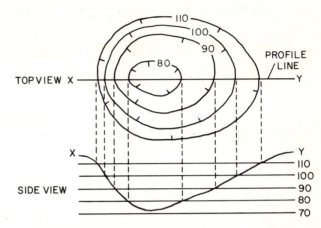

**Figure 7-3.** Map view and topographic profile of a depression.

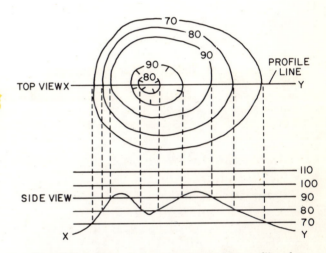

**Figure 7-4.** Map view and topographic profile of a hill and depression.

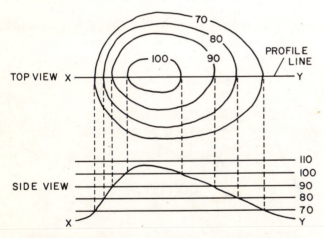

**Figure 7-2.** Map view and topographic profile of a hill.

The vertical distance between contour lines is called the ==contour interval== and is indicated on all topographic maps. Selection of contour interval is a function of map scale and the variation of elevation within an area. On most maps, every 50 or 100 foot ==contour line is indexed by marking it a darker color.==

A few simple rules to follow when reading or constructing topographic maps are listed below:

1. ==C==ontour lines connect points of equal elevation.

2. ==C==ontour lines do not cross or divide each other.

3. ==T==he spacing of contour lines reflects the "steepness" of a slope; wide spacing for gentle slopes, close spacing for steep slopes (see Figures 7-2, 7-3, 7-4).

4. ==C==ontour lines which cross stream valleys make a "V" which points in the upstream direction.

5. ==E==very topographic map has a specific contour interval.

Topographic maps may be used to quantitatively express specific topographic features. For example, the difference in elevation between two points (called the *relief*) may be measured. In addition, the *total* (or *maximum*) *relief* within an area (maximum difference in elevation) may be computed. Topographic maps also allow for the *gradient or slope* of a surface to be measured. Most often the gradient is expressed as a change in vertical height (measured in feet) compared to the change in horizontal distance (measured in miles). Measuring the gradient requires the accurate determination of point elevation and effective use of a map scale. For example, two points are at elevations 182 and 118 feet respectively. They are separated by a horizontal distance of 1.8 miles on a topographic map. The gradient (or slope) between the two points may be calculated as follows:

$$\text{gradient (ft/mi)} = \frac{\text{difference in elevation (ft)}}{\text{horizontal map distance (mi)}}$$

$$= \frac{64 \text{ ft}}{1.8 \text{ mi}}$$

$$= 35 \text{ ft/mi}$$

Stream and river gradients may be measured in a similar fashion. However, horizontal map distances must be measured along channels. This can best be accomplished by noting where a stream or river crosses two contour lines and dividing the channel into a series of nearly straight-line segments between the contours. The gradient would be the sum total of the lengths of these individual segments divided into the difference in elevation between the two contour lines (Fig. 7-5). Always remember that the gradient you calculate in this manner is an *average* between two points. The actual ground surface may be highly irregular between the points.

To better visualize the surface topography of an area, it is often advantageous to construct a *topographic profile.* The construction of a topographic profile is not difficult if the following procedure is used (examine examples illustrated in Figures 7-2, 7-3, 7-4).

1. Place a strip of paper along a selected profile line.
2. Mark and label on the paper the exact place where each contour line, stream valley and hill crest crosses the profile line.
3. Construct a vertical grid scale.
4. Place the paper strip above the grid and project marked contour positions onto the vertical grid scale at the proper elevations.
5. Connect all points on the grid with a smooth even line which is consistent with the topography.

The resultant profile is essentially a silhouette of the topography when viewed at right angles to the profile line.

The horizontal scale of a topographic profile is identical to that of the map from which it is constructed. However in most instances, if the ver-

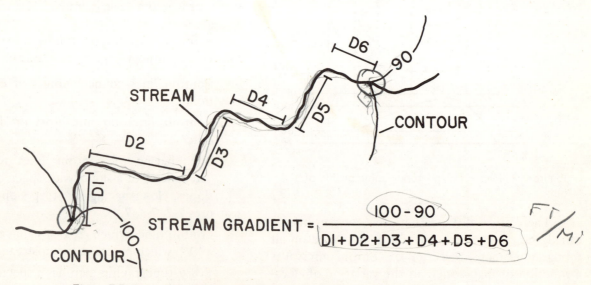

**Figure 7-5.** Diagram illustrating how stream gradients may be calculated.

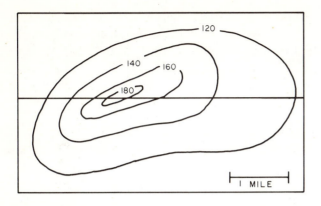

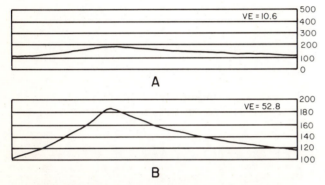

**Figure 7-6.** Diagrams illustrating how changing vertical exaggeration affects the detail of topographic profiles. Note how increasing the vertical exaggeration from 10.6 (in profile A) to 52.8 (in profile B) permits a better view of the topography shown in the map.

tical scale of a profile is the same as the horizontal scale there would be very little variation of elevation indicated. To emphasize the shape of the land surface, most topographic profiles have some *vertical exaggeration.* Vertical exaggeration is produced by increasing the vertical scale of a topographic profile so that surface irregularities may be adequately portrayed. The amount of vertical exaggeration should always be noted on the profile. It may be calculated by dividing the vertical scale of the profile grid into the horizontal (map) scale using the *same* units of distance and elevation. For example, if a vertical scale of 1 inch = 100 feet is used (this would mean that 1 inch on the grid equals 1,200 inches of elevation) and the map scale is 1 inch = 1 mile (1 inch on the map is equal to 63,360 inches of horizontal distance on the map), the vertical exaggeration would be 63,360/1,200 or 52.8. This indicates that vertical distances shown on the profile are 52.8 times greater than the actual distances. The effect of changing the vertical exaggeration is illustrated in Figure 7-6.

f. What is the contour interval of the map? , 2 0

g. What is the precise location of the University of Georgia Coliseum using the latitude and longitude system? LAT 83° 22' 56 15"    LONG 33° 58' 15"

h. What is the location of the University of Georgia Coliseum using the rectangular coordinate system?

i. What is the elevation above sea level of the University of Georgia Coliseum? 690 FT

j. What is the gradient of the Middle Oconee River between just south of US Route 129-441 and just north of US Route 29 (by the Athens Corporate Boundary line)? 4 FT/MI

k. Construct a topographic profile along a line which connects the University of Georgia Coliseum and the Friendship Church. Use a vertical exaggeration of 50X in your profile.

# Exercise 8

# Stream Patterns

If precipitation exceeds the amount of moisture which can be absorbed into the ground, water accumulates on the land surface. Because of the influence of gravity it will begin to flow downhill, being channeled into streams of increasing size. The number of streams within a specific area is termed the *drainage density*. It is largely controlled by climate and the porosity and permeability of the bedrock. Rocks with low porosity and permeability (such as shale, slate, and more crystalline igneous rocks) can absorb less precipitation than rocks with high porosity and permeability (such as sandstone). As a result, a greater number of streams will develop in regions underlain by rocks with low porosity and permeability.

The geometric relationship of streams within an area is termed the *drainage pattern*. It is controlled by several factors, including relief, type of bedrock, and structural features (faults, folds, joints, etc.). Although there is considerable variability in these factors, the actual number of natural drainage patterns is limited. The most common type of pattern resembles the branching arms of a tree and is termed *dendritic* (Fig. 8-1A). This type of pattern commonly develops in areas where there is little variation in the type of bedrock and no structural control on stream location. A more angular type of drainage pattern is typically developed in areas where intersecting faults or joints control stream location. It is characterized by streams with abundant right angle bends and is named *rectangular* (Fig. 8-1B). Another angular type of drainage pattern is typical of areas where structural features or differences in bedrock have directed streams to a single major trend. Smaller tributaries generally converge at nearly right angles to this trend and form what is termed a *trellis* drainage network (Fig. 8-1C). In topographically high areas drained by streams which flow downhill in all directions, a fourth type of drainage pattern is commonly developed. The pattern resembles spokes of a wheel and is termed *radial* (Fig. 8-1D). These four major types of drainage patterns are ideal end-members and may be combined in any specific region. For example, it is very common to find a regional rectangular drainage pattern locally modified by dendritic tributaries (Fig. 8-1E).

To better visualize the drainage system of an area, it is often helpful to construct a *drainage map*. This consists of tracing all drainage features on a transparent overlay.

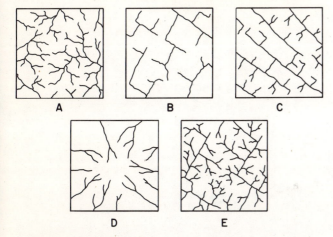

**Figure 8-1.** Common types of drainage patterns: (A) dendritic, (B) rectangular, (C) trellis, (D) radial, (E) composite (overall rectangular, locally dendritic).

## EXERCISE 8

This laboratory will introduce the study of drainage patterns. Following this exercise you should be able to:

1. Recognize the major types of drainage patterns.
2. Evaluate the various factors which control drainage patterns and drainage density.

## QUESTIONS

1. Atlanta, Georgia-Alabama 1:250,000 map (supplied in laboratory).

    a. Prepare a drainage map for the area between 33°00′ and 33°15′ latitude and 84°15′ and 84°45′ longitude.

    b. How many square miles does this area represent?

    c. What type of drainage pattern is present?

    d. What does the drainage pattern indicate about the variability of bedrock within the area?

2. Charlottesville, Virginia-West Virginia 1:250,000 map (supplied in laboratory).
    a. Prepare a drainage map on a tracing paper overlay for the area between 38°30′ and 38°45′ latitude and 79°15′ and 79°45′ longitude.

    b. How many square miles does this area represent?

    c. What type of drainage pattern is present?

    d. How does the drainage density in this area compare to that in question 1?

    e. What can you say about differences in the porosity and permeability of the bedrock in these two areas?

    f. What explanation can you give for the different drainage patterns in the two areas?

3. Katahdin, Maine 15′ Quadrangle (supplied in laboratory).

    a. Prepare a drainage map for the area between 43°05′ and 43°07′30″ latitude and 77°00′ and 77°05′ longitude.

b.   How many square miles does this area represent?

c.   What type of drainage pattern is present?

d.   What appears to be the major control on the drainage pattern developed in this area?

4. Kaaterskill, New York area (map in manual; Fig. 8-2).

a.   Prepare a drainage map for this area.

b.   How many square miles does this area represent?

c.   What type(s) of drainage pattern(s) are present?

d.   Are there any major differences in drainage density within this area?

e.   The area is mostly underlain by sandstones and shales. On the basis of your drainage map, where do you think these two rock types are found? Indicate the probable distribution of these rocks on your drainage map. On what criteria do you base your answer?

f.   Is the distribution of the two rock types reflected in any other ways?

5.  Athens West, Georgia 7 1/2′ Quadrangle (supplied in laboratory).

a.   Prepare a drainage map for the SC and SE rectangles of the map (use two pieces of tracing paper).

b.   How many square miles does this area represent?

c.   What type of drainage pattern is present?

d.   What can you say about variations in the type of bedrock in this area?

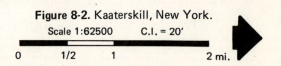

**Figure 8-2.** Kaaterskill, New York.

Scale 1:62500      C.I. = 20′

0        1/2        1                    2 mi.

# Exercise 9

# Stream Erosion

Running water has considerable energy to transport rock and mineral fragments produced by weathering. It is an important erosional agent and plays a significant role in sculpturing the earth's surface. Development of a regional landscape through stream erosion is not a chance or random process, but occurs through a series of generally predictable stages (illustrated in Fig. 9-1). This

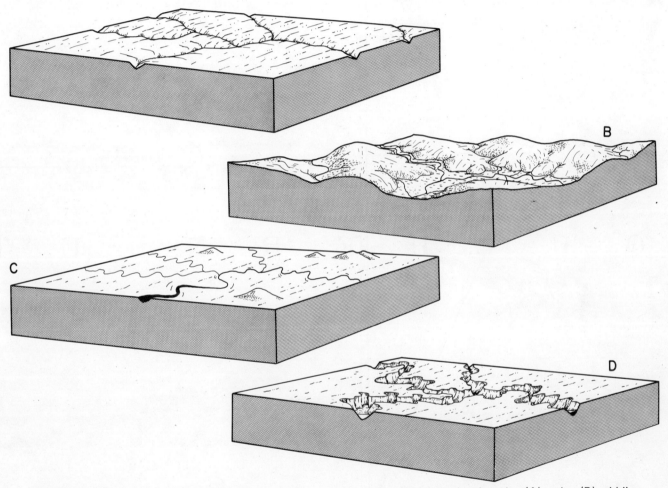

**Figure 9-1.** Generalized diagrams illustrating various stages in the fluvial erosional cycle: (A) early, (B) middle, (C) late, (D) rejuvenation.

general progression of landscape development is termed the *fluvial erosion cycle.* The final result of the cycle is controlled by the *base level* (the lowest point to which streams can flow). In regions with well-integrated and through-flowing drainage networks, base level approximately coincides with sea level. Stream erosion will attempt to reduce the landscape to this elevation. In areas which are internally drained, local base levels (generally at higher elevations than sea level) control the lowest point to which erosion may progress. For example, streams draining the terrane in the vicinity of Lake Superior are attempting to erode the landscape to the elevation of the lake.

In *early* or *youthful* stages of the erosional cycle, landscapes are relatively flat and only a small portion of the land surface is in slopes (Fig. 9-1A). Drainage is generally poor (low drainage density). Streams have relatively steep gradients and are actively eroding downward. They have little or no flood plains and commonly flow in straight, narrow, steep-sided valleys with "V"-shaped profiles.

An example of a region in an early stage of stream erosion is the Soda Canyon, Colorado area (Fig. 9-2). Note that most of the streams in this region occupy steep-sided canyons while interstream areas are generally flat and undissected. The percentage of the land surface in slope is small. The overall drainage is not well-integrated and there is a low drainage density. The accompanying topographic profile (Fig. 9-3) shows the marked "V"-shaped valley profile which is characteristic of regions in the early stages of stream erosion.

If stream erosion proceeds, landscapes pass into a *middle* or *mature* stage of development (Fig. 9-1B). Interstream areas are progressively eroded back and stream valleys become wider. As a result, a greater portion of the landscape is in slope. Drainage systems become better developed with a greater drainage density. Major streams have gentler gradients and lateral erosion dominates over downcutting. Streams begin to meander and have less pronounced "V"-shapes than in the early stages of erosion.

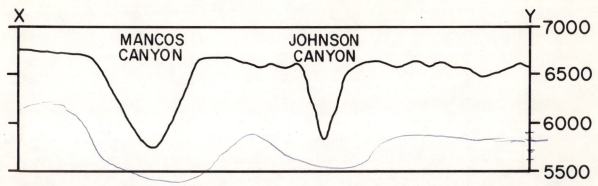

**Figure 9-3.** Topographic profile along line X-Y in the Soda Canyon, Colorado area (Fig. 9-2); vertical exaggeration = 5.2X.

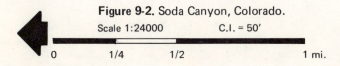

**Figure 9-2.** Soda Canyon, Colorado.
Scale 1:24000          C.I. = 50'

0          1/4          1/2                    1 mi.

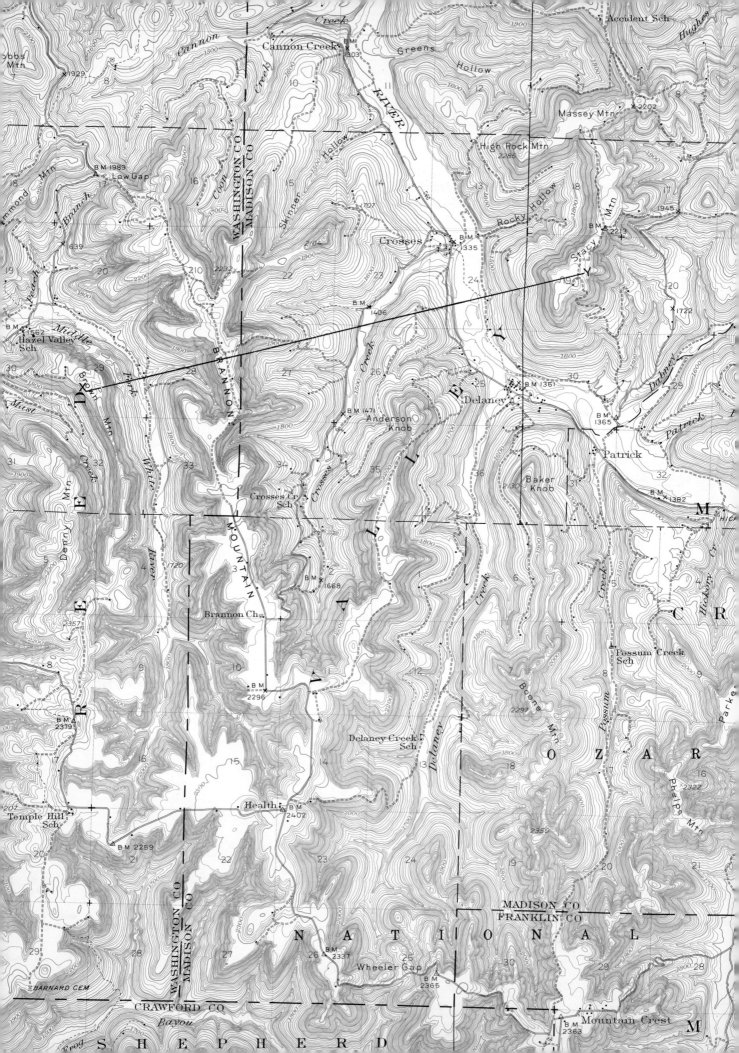

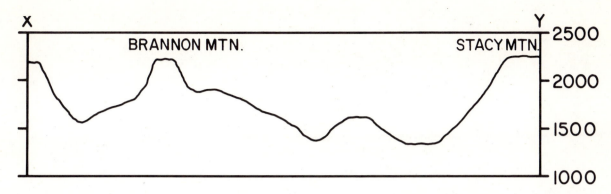

**Figure 9-5.** Topographic profile along line X-Y in the St. Paul, Arkansas area (Fig. 9-4); vertical exaggeration = 5.2X.

The St. Paul, Arkansas region (Fig. 9-4) is an area in the beginning stages of middle erosional development. Note that more of the area is in slope than in the Soda Canyon, Colorado area. Stream density is increased and the larger streams meander over restricted flood plains. The accompanying topographic profile (Fig. 9-5) shows that stream valleys have a more flattened "V"-shape than the Soda Canyon area. Also note that although stream valleys have been widened, interstream areas are still flat and have generally similar elevations. They are the remants of an older plateau similar to that present throughout most of the Soda Canyon area.

The Hazard South, Kentucky region (Fig. 9-6) typifies an area in a more advanced stage of middle erosional development. Essentially all of the area is in slope. Interstream areas have been reduced to narrow ridges with rounded summits. No remnants of an older plateau remain. The area has a well-integrated drainage system. Large rivers and smaller tributaries form a dendritic drainage pattern and meander over variably developed flood plains. The accompanying topographic profile (Fig. 9-7) shows the flattened "V"-shaped valley profiles which are typical of this stage of the fluvial erosional cycle.

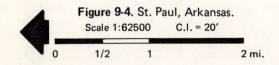

**Figure 9-4.** St. Paul, Arkansas.
Scale 1:62500      C.I. = 20'

0      1/2      1              2 mi.

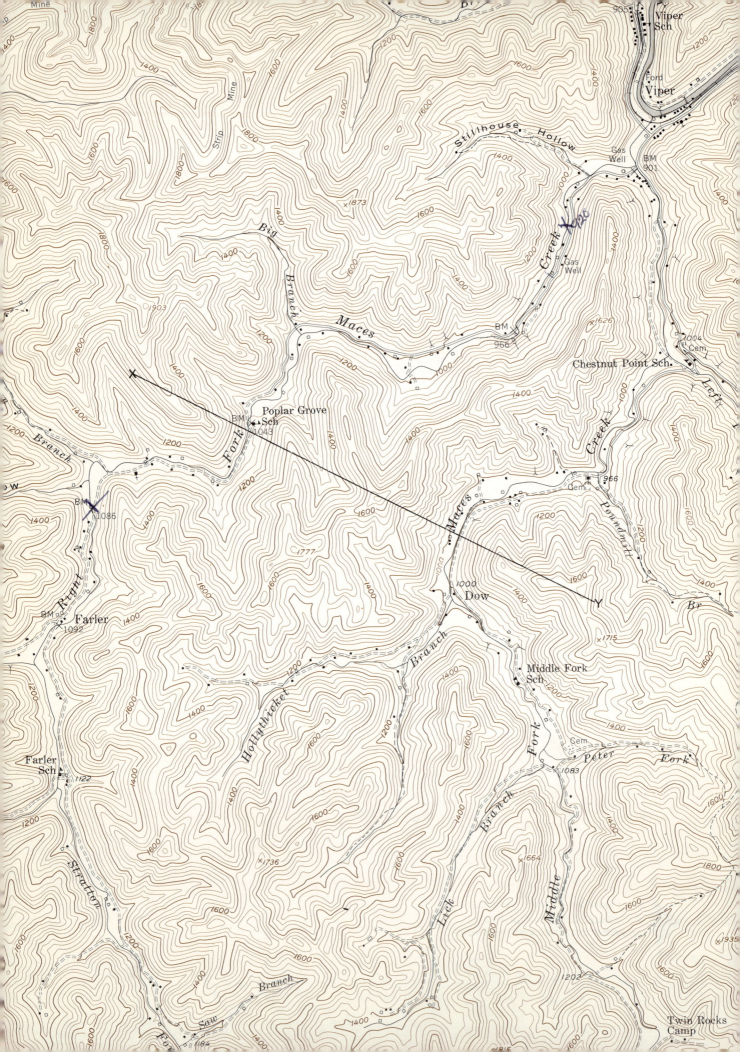

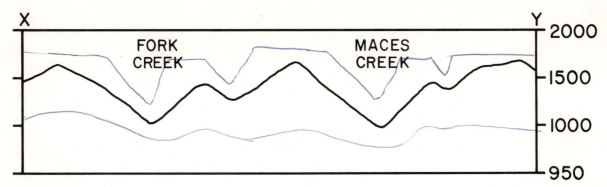

**Figure 9-7.** Topographic profile along line X-Y in the Hazard South, Kentucky area (Fig. 9-6); vertical exaggeration = 2X.

If base level remains constant for a prolonged period of time, extensive erosion may occur and form landscapes in a *late* or *old age* stage of development (Fig. 9-1C). Regions in this erosional stage are usually poorly drained and contain abundant swamps or marshes. Most of the landscape is flat and interstream areas are no more than broad, low hills. Streams have very low gradients and are actively depositing material rather than eroding downward or laterally. They freely meander over large flood plains. Oxbow lakes are common. The flat landscape is nearly at base level elevation and is termed a *peneplain.* Rocks which are very resistant to erosion may, locally, remain as isolated hills which are termed *monadnocks.*

The Campti, Louisiana area (Fig. 9-8) is typical of a region in the late stage of the fluvial erosional cycle. Overall drainage is not well-integrated and there are abundant marshy and swampy areas. Note that the Red River meanders over an extensive flood plain. The accompanying topographic profile (Fig. 9-9) illustrates the low relief of the area.

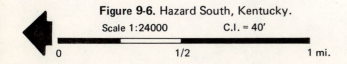

**Figure 9-6.** Hazard South, Kentucky.
Scale 1:24000          C.I. = 40'

0                    1/2                    1 mi.

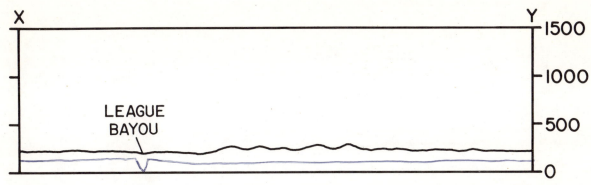

**Figure 9-9.** Topographic profile along line X-Y in the Campti, Louisiana area (Fig. 9-8); vertical exaggeration = 5.2X.

At any time during the fluvial cycle of erosion a region may be tectonically uplifted, thereby lowering the base level to which streams flow. As a result, streams begin more rapid downcutting and erosional characteristics are moved back toward those which typify a youthful area. This process is termed *rejuvenation* (Fig. 9-1D). Landscapes which result from this process are generally complex, as remnants of older landforms are often preserved between areas where active downward erosion is occurring. For example, remnants of old flood plains are commonly preserved as *terraces* along the margins of more actively downcutting streams. Also, streams which had meandering courses prior to rejuvenation often begin active downcutting in the same channels and thereby become entrenched into the old meanders. This results in the anomaly of meandering streams (typical of a middle or late erosional stage) which have "V"-shaped valley profiles (typical of an early erosional stage). These are termed *entrenched meanders* and are most distinctive of rejuvenated areas.

The Renovo West, Pennsylvania area (Fig. 9-10) is an example of a rejuvenated terrane. Note that both major rivers and smaller tributaries meander but have steep-sided, canyonlike valleys similar to those seen in the Soda Canyon, Colorado area (Fig. 9-2). The accompanying topographic profile

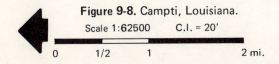

**Figure 9-8.** Campti, Louisiana.
Scale 1:62500    C.I. = 20′

0    1/2    1    2 mi.

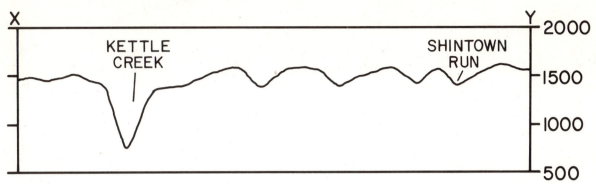

**Figure 9-11.** Topographic profile along line X-Y in the Renovo West, Pennsylvania area (Fig. 9-10); vertical exaggeration = 5.2X.

**TABLE 9-1**

Characteristics of the Various Stages of the Fluvial Erosional Cycle

| Stage | Valley Profile | Interstream Areas | Percentage of Area in Slope | Other Features |
|---|---|---|---|---|
| Early | | broad, flat | small | straight stream courses, down-cutting, waterfalls, rapids |
| Middle | | narrow, variably rounded | large | lateral erosion, larger streams may meander, small flood plains |
| Late | | broad, flat | small | deposition, meanders, swamps, oxbow lakes, extensive flood plains |
| Rejuvenation | | broad, variable topography, often composite | variable | downcutting, entrenched meanders, terraces |

(Fig. 9-11) shows that the "V"-shape of the stream valleys is typical of a youthful stage of erosion. In contrast to the Soda Canyon area, the interstream areas of the Renovo West area are not flat undissected plateaus, but have a topography which is typical of a previous middle stage of the erosional cycle.

As pointed out earlier, the ultimate goal of stream erosion is to reduce a landscape to a generally flat surface with an elevation approximate to that of the local base level. This is accomplished in a series of stages as stream valleys are widened and weathering products transported out of the area. This overall process is illustrated in Figure 9-12 by a comparison of topographic profiles which are common to the various stages of the erosion cycle. Table 9-1 lists the distinguishing characteristics of

each stage of the fluvial erosion cycle. It should always be remembered that the various erosional stages are completely gradational. Even an individual stream will commonly show different stages of landscape development along its course.

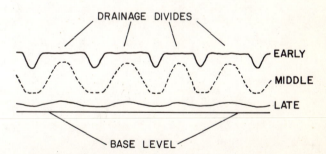

**Figure 9-12.** Diagrammatic sketch of generalized topographic profiles which characterize various stages of the fluvial erosional cycle.

**Figure 9-10.** Renovo West, Pennsylvania.

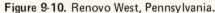

Scale 1:62500     C.I. = 40'

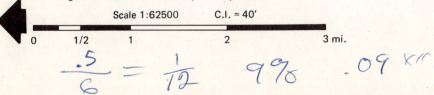

0     1/2     1     2     3 mi.

$$\frac{.5}{6} = \frac{1}{12} \quad 9\% \quad .09 \text{ km}$$

## EXERCISE 9

This laboratory exercise consists of examining a suite of topographic maps which illustrate landforms which are characteristic of the various stages in the fluvial erosional cycle. It is designed both to further develop your skills in working with topographic maps and to foster an awareness of the landscapes around you and how they formed.

When working with the maps always be certain to first check the scale and contour interval to give a perspective of the landforms which you will study. An ample supply of graph paper is provided for use in construction of topographic profiles. Work in pencil and please do *not* mark on the maps which are supplied in the laboratory. Locations of all points discussed are given by the rectangular coordinate system.

At the completion of this exercise you should be able to:

1. Identify landscapes which are a result of stream erosion.
2. Recognize the erosional stage of a specific area.
3. Predict what effects continued stream erosion will have on a landscape.

## QUESTIONS

1. Soda Canyon, Colorado area (map in manual; Fig. 9-2).

   a. This area is in an early stage of stream erosion. List the most distinguishing features of this erosional stage which are illustrated in this area.
   FLAT
   LO % IN SLOPE            RIVER CONFINED TO BOTTOM OF VALLEY
   HI GRADIENT                 ( LITTLE FLOOD PLAIN)
   DEEP U SHAPED VALLEYS (STEEP SIDED)

   b. Calculate what percentage of the area is in slope along the profile line.
   22.7%

   c. What is the maximum relief along the profile line?
   950 - 6750 TO 5800

   d. What is the average gradient (in ft/mi) of the stream in Johnson Canyon (between Mancos Spring and the Mancos River)? 6686 - 5700 OVER 2.62 MI
   376.3 FT/MI

   e. If erosion proceeds, this area will progress into a middle stage of erosion. On the topographic profile (Fig. 9-3), sketch a hypothetical profile which would characterize a middle erosional stage landscape along the profile line.

2. Hazard South, Kentucky area (map in manual; Fig. 9-6).

   a. This area is in a middle stage of stream erosion. List the most distinguishing features of this erosional stage which are illustrated in this area.
   NEARLY All AREAS IN SLOPE
   BROADER VALLEY - RIVER MEANDERS SLIGHTLY
   LOWER GRADIENT

   b. Calculate what percentage of this area is in slope along the profile line.
   100%

   c. What is the maximum relief along the profile line?
   680 - 1680 TO 1000

d.   What is the average gradient (in ft/mi) of Maces Creek (between Stillhouse Hollow and Wooten Branch Creek)?   *1086 - 920 over 2.875mi*
*57.74 FT/mi*

e.   Using a dashed line, sketch on the topographic profile (Fig. 9-7) how this area might have appeared in an earlier erosional stage. Using a dotted line, indicate how the area might appear in the future if erosion continues.

3.   Campti, Louisiana area (map in manual; Fig. 9-8).

a.   This area is in a late stage of stream erosion. List the most distinguishing features of this erosional stage which are illustrated in the area.

*POOR DRAINAGE      MEANDERING STREAMS*
*SWAMPS                    OXBOW LAKES*
*FLAT                           POINT BARS*
*BROAD VALLEYS        CUTBANKS*

b.   Calculate what percentage of the area is in slope along the profile line.
*54.5%*

c.   What is the maximum relief along the profile line?
*120 — 220 to 100*

d.   What is the approximate gradient of the Red River in this area?
*essentially 0*

e.   If this area were tectonically uplifted, it would undergo a process of rejuvenation. Using a dashed line, indicate on the topographic profile (Fig. 9-9) how this area might appear if rejuvenation were to occur.

4.   Fill in the following table with your numerical answers to questions 1-3. Discuss which features may be used as a diagnostic index of erosional stage.

Characteristics

| Erosional Stage | Stream Gradient | Maximum Relief | Percentages of Area in Slope |
|---|---|---|---|
| early | | | |
| middle | | | |
| late | | | |

*SAINT PAUL ARKANSAS*
*29.0% IN SLOPE*

5.   Lake McBride, Kansas 7 1/2′ Quadrangle (supplied in laboratory).

a.   Construct a topographic profile along a line connecting the 3069′ elevation at the intersection of the two roads (C-WC) and the "V" in Beaver (NE-SC). Use a vertical scale of 1″ = 100′. What is the vertical exaggeration of your profile?

b.　What percentage of the area is in slope along the profile line?

c.　What is the maximum relief along the profile line?

d.　In what stage of erosion is this area? On what criteria do you base your answer?

6.　Refuge, Askansas-Mississippi 15′ Quadrangle (supplied in laboratory).

a.　What is the maximum relief in this area?

b.　In what stage of erosion is this area? On what criteria do you base your answer?

c.　What type of geomorphic feature is Lake Chicot?

7.　Roman Nose Mountain, Oregon 15′ Quadrangle (supplied in laboratory).

a.　Construct a topographic profile along a line connecting Esmond Lake (EC-C) and BM1537 (EC-SC). Use a vertical scale of 1″ = 1,000′.

b.　In what stage of erosion is this area? On what criteria do you base your answer?

c.　Using a dashed line, indicate on your topographic profile how this area might have looked in an earlier erosional stage. Using a dotted line, indicate how it might appear in the future.

8.　Oolitic, Indiana 15′ Quadrangle (supplied in laboratory).

a.　Construct a topographic profile along a line connecting the "M" in Marion (C-SE) and the town of Croxton (NW-SE). Use a vertical scale of 1″ = 100′. What is the vertical exaggeration of your profile?

b.　What type of geomorphic feature is developed along the East Fork of the White River in your profile?

c.　What can you say about the erosional history of this area?

9.　Stone Mountain, Georgia 7 1/2′ Quadrangle (supplied in laboratory).

a.　Construct a topographic profile along a line connecting Lake Arrowhead (EC-SC) and the "i" in Mountain of Stone Mountain Creek (EC-EC). Use a vertical scale of 1″ = 200′. What is the vertical exaggeration of your profile?

77

b. What is the maximum relief along the profile line?

c. What is the maximum relief along the profile line if you do not consider Stone Mountain?

d. Stone Mountain is underlain by a resistant type of rock called quartz monzonite (similar to granite). The remainder of the area is underlain by relatively nonresistant metamorphic rocks. Given this information, how do you explain the anomalous relief of Stone Mountain?

e. What type of geomorphic landform is Stone Mountain?

f. What type of geomorphic landform is the remainder of the area?

g. In what stage of erosion is this area? On what criteria do you base your answer?

10. Athens West, Georgia 7 1/2' Quadrangle (supplied in laboratory).

Examine the map and the topographic profile which you constructed in Exercise 7. In what stage of erosion is this area? On what criteria do you base your answer?

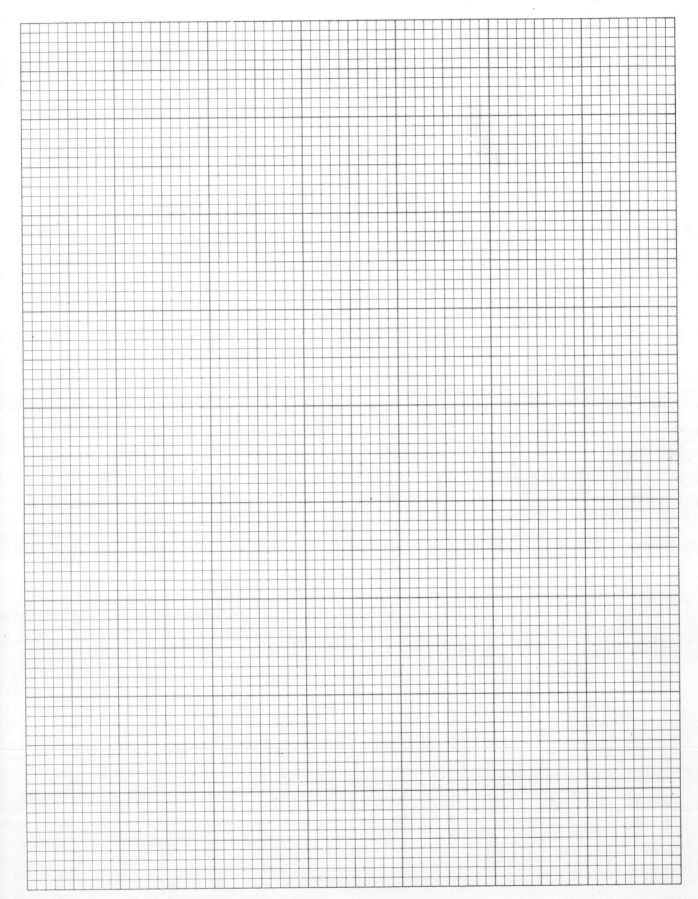

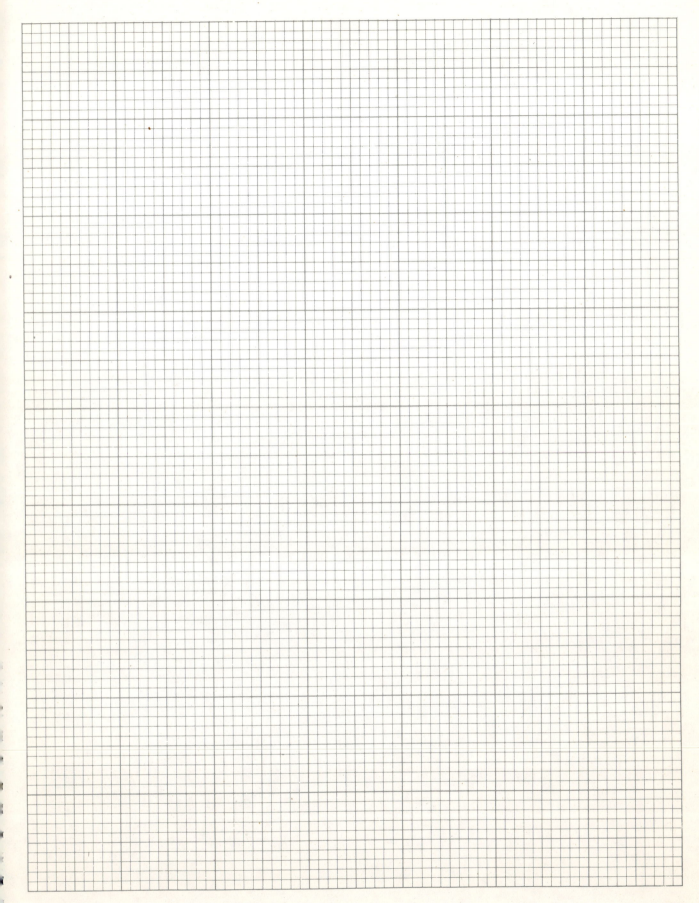

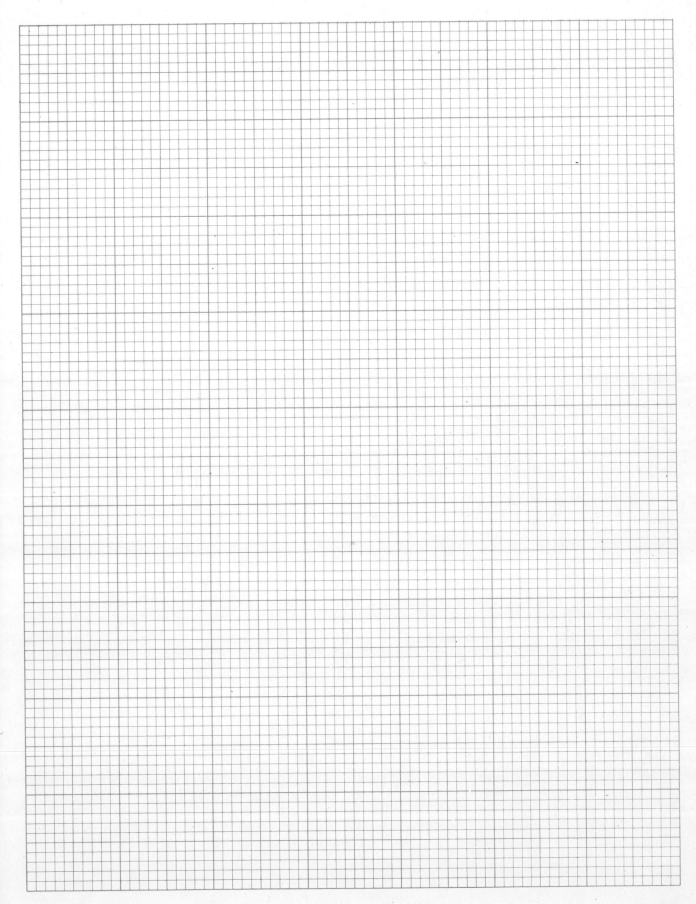

7920

# Exercise 10

# Ground Water Erosion

Precipitation which is absorbed into the ground is termed *groundwater.* It migrates downward through soil and rock until it reaches a level below which open spaces and voids are filled with water. This level marks the upper surface of the *water table.* The position of the water table is variable and generally parallels overall irregularities in surface topography; being slightly higher under hills and slightly lower under valleys. Where the water table intersects the land surface, springs, lakes, and streams may occur.

Water which has percolated downward to the water table does not remain in a fixed position. It slowly migrates through rock and soil down the gradient of the water table. During groundwater movement erosion may occur. Areas which are underlain by carbonate rocks (limestone and dolomite) are particularily susceptible to groundwater erosion because of the solubility of calcium carbonate minerals.

The solution activity of groundwater leads to the widening of existing joints and fractures in carbonate rocks and underground caves or caverns may develop. Formation of these underground features is eventually reflected in the surface landscape as circular depressions or *sinks* often form by collapse of cave or cavern roofs. If sinks intersect the water table they fill with water and form circular lakes. In more advanced stages of solution erosion, underground cavern systems enlarge and surface drainage combines with the groundwater system. Sinks may coalesce and form *solution valleys.* Surface drainage is disrupted so that only major streams flow in well-defined valleys. Tributaries are few and may suddenly disappear in sinks or solution valleys. Landscapes which are characterized by these solution features have what is termed a *karst* topography.

The Interlachen, Florida area (Fig. 10-3) is underlain by limestone and has a well-developed karst topography. Note the abundant depressions or sinks which evidence the extensive subsurface solution activity of groundwater. Many of the sinks are filled with water indicating that they are interconnected with the water table. The water elevation in each sink is a record of the level of the water table at that point and the lakes may be used to outline subsurface variations in the water table level. For example, the accompanying topographic profile (Fig. 10-1) is constructed along line X-Y on the Interlachen map. The dashed line connects elevations of the lakes along the profile line and, therefore, indicates the variation in water table level. Note that the water table is sloping toward the southeast. This suggests that groundwater will migrate in that general direction. It is also possible to contour water table levels by using lake eleva-

tions in a procedure similar to that of using surface elevations to construct a topographic map (Fig. 10-2).

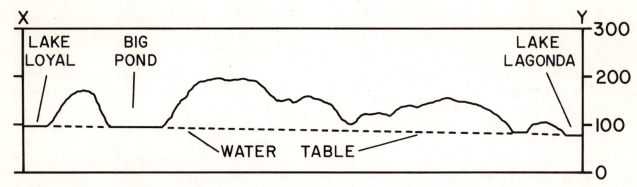

**Figure 10-1.** Topographic profile along the line X-Y on the Interlachen, Florida map (Fig. 10-3); vertical exaggeration = 26X. Dashed line indicates variation in depth of water table along the profile line.

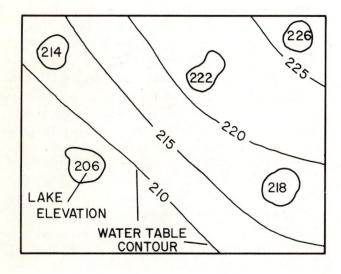

**Figure 10-2.** Diagram illustrating how lake elevations may be used to contour water table levels.

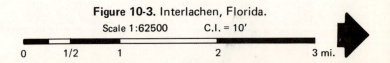

**Figure 10-3.** Interlachen, Florida.
Scale 1:62500     C.I. = 10'

# EXERCISE 10

In this exercise the study of landforms is continued. Landscapes which have been produced largely by the erosional activity of groundwater will be examined and compared to those produced by stream erosion. After completion of this exercise you should be able to:

1. Recognize landscapes which have formed by the solution activity of groundwater.
2. Estimate subsurface variations in water table levels.

## QUESTIONS

1. Interlachen, Florida area (map in manual; Fig. 10-3).

   a. On a tracing paper overlay, outline all major lakes in the area and note their elevations. Using the procedure illustrated in Figure 10-3, construct a contour map of the upper surface of the water table in this area.

   b. In what general direction will groundwater migrate in the area? What is the average gradient of the water table? *EAST*

   c. Approximately 2 miles west of Interlachen on Route 20 is a 131 foot Bench Mark notation. If you were to build a house at this point, how much would it cost you to have a well drilled to the water table (the going rate is around $15/foot)? *WATER TABLE AT ABT 85FT 46FT Hole $690.00*

2. Mammoth Cave, Kentucky 15' Quadrangle (supplied in laboratory).

   This area is underlain by a resistant sandstone unit (capping the high areas) and a less resistant limestone unit (underlying the low areas).

   *FOR THEM*

   a. Construct a topographic profile along a line connecting Turnhole Ferry (SW-NC) and the town of Rock Hill (NW-SC). Use a vertical scale of 1″ = 300′. What is the vertical exaggeration of your profile?

   b. Knowing that the higher elevations are held up by a more resistant sandstone unit and that the lower elevations are underlain by limestone, add a generalized contact between the two units on your profile.

   *ASKTHEM*

   c. How does the distribution of these two rock types serve to explain the variation of landscape in this area?

   d. What are the numerous depressions in the southern half of the area? How did they form?

   e. Examine Sinking Creek (S-SC). In what direction is it flowing?

f. What happens to this creek north of Shively School?

g. Name several other streams in the southern half of the area which are similar to Sinking Creek.

3. Crystal Lake, Florida 7 1/2′ Quadrangle (supplied in laboratory).

This area is in the heart of the Florida citrus region. It is underlain by Eocene limestone. A citrus processing company plans on building a new plant in the vicinity of Blue Pond (SW-NW). They are going to pump waste effluents into the pond. They are also going to build a preliminary pulp reduction facility near the road junction immediately northeast of Long Lake (EC-NE). They plan on using that lake for waste storage. Due to the lack of housing in this area, the company is going to build living quarters for their employees on Route 52 near Piney Pond (SE-NC). They plan on obtaining drinking water from the immediate vicinity of Piney Pond.

Your job is to write a brief environmental impact statement discussing the potentially harmful effects of the proposed construction. Preparation of several topographic profiles (using a vertical scale of 1″ = 50′) will help evaluate possible problems. What changes would you suggest to the company?

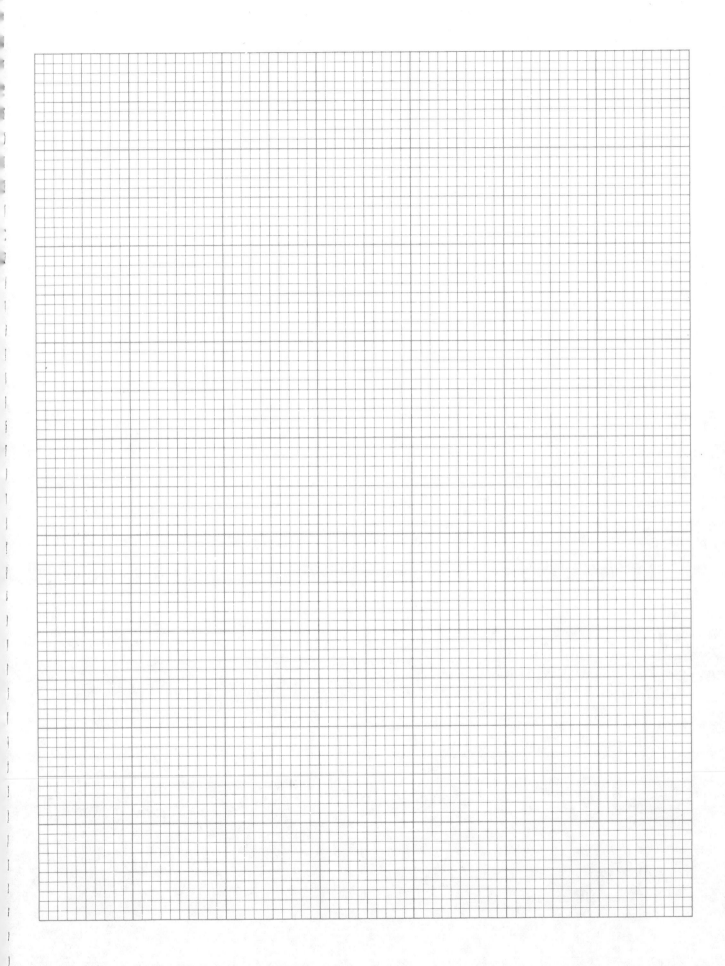

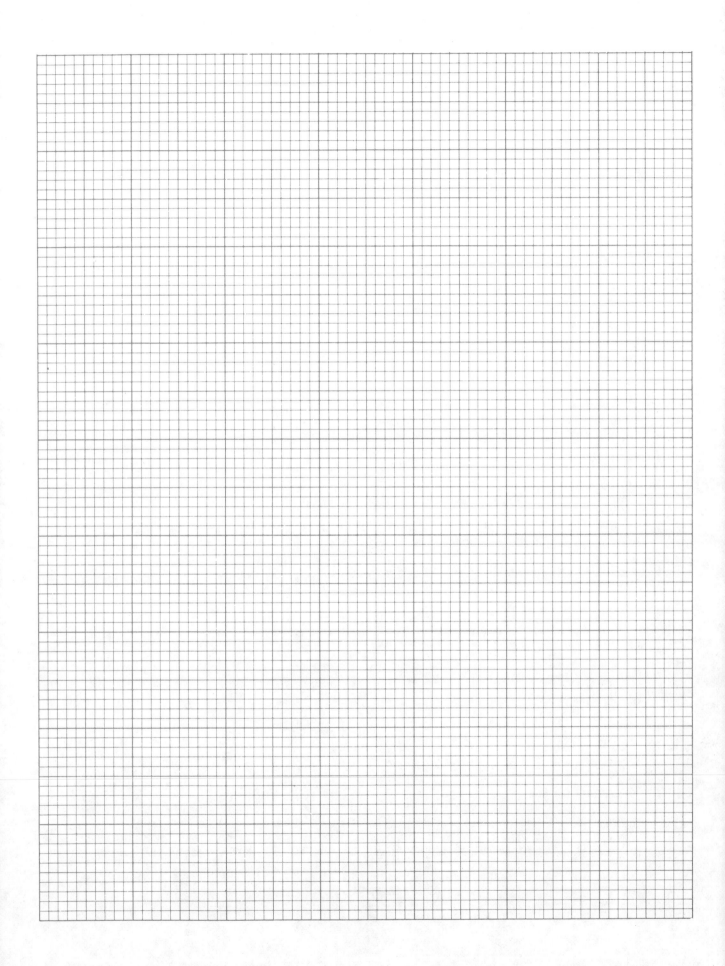

# Glacial Erosion

Thick sheets of slowly moving ice (*glaciers*) are very effective erosional and depositional agents. Although there have been only a few periods of glaciation in the earth's history, one of the major glacial episodes has just ended or is still in progress. As a result, landforms which presently characterize large portions of the earth's surface are a product of the work of ice. Glaciers may be classified into two types: *continental* and *alpine* (or *mountain*). Continental glaciers are extensive ice sheets such as those which cover most of Greenland and Antarctica today. Alpine glaciers are more restricted in extent, being confined to areas of high elevations (many are found today in mountainous areas). Both types of glaciers produce landforms which are readily identified on topographic maps.

## Continental Glaciers

Continental glaciers are powerful erosional agents and act as giant bulldozers, effectively stripping off soil and partially weathered bedrock as they advance. This scouring tends to streamline and subdue pre-existing landforms. It also disturbs pre-existing drainage systems. When a continental glacier melts and retreats, the debris which was picked up and incorporated in the ice is deposited and forms distinctive landforms. The more common depositional features are listed below and illustrated in Figure 11-1. The Whitewater, Wisconsin (Fig. 11-2), Palmyra, New York (Fig. 11-3), and Passadumkeag, Maine (Fig. 11-4) maps show the characteristic expression of many of these depositional features.

1. *Terminal Moraines.* Erosional debris carried by a glacier is chaotically dropped during melting at its terminus. This material piles up in front of the glacier and forms a linear ridge which generally follows the margin of the ice sheet. This ridge is called a *terminal moraine.* Small stranded blocks of ice are commonly buried by the erosional debris and when they eventually melt, small depressions or kettles form. As a result, terminal moraines usually have a highly irregular topography characterized by abundant knobs and depressions. The low ridge which trends northeast-southwest across the central portion of the Whitewater, Wisconsin map (Fig. 11-2) is a good example of a partially dissected terminal moraine deposit.

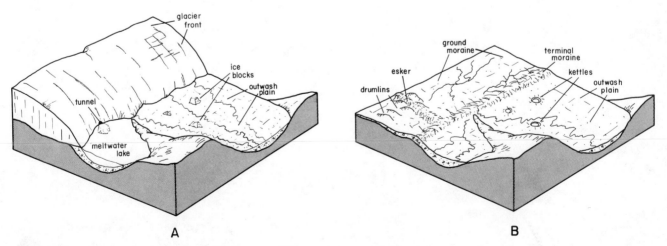

**Figure 11-1.** Generalized diagrams showing the development of depositional landforms which result from continental glaciation: (A) prior to glacial retreat; (B) after glacial retreat.

2. *Outwash Plains.* Large quantities of water result from the melting of a glacier. This meltwater flows outward from the terminus in a vast array of small streams which carry a considerable amount of fine-grained erosional material. This debris is deposited on a broad *outwash plain* in front of the terminal moraine. This plain is generally flat except for isolated kettles (now often small lakes or marshy areas). The area southeast of the terminal moraine in the Whitewater, Wisconsin map (Fig. 11-2) is a partially dissected outwash plain.

3. *Ground Moraine.* When a continental glacier begins to actively recede, erosional material which it has incorporated is deposited as a sedimentary blanket (or *ground moraine*) behind the terminal moraine. This regional sedimentary blanket totally disrupts pre-existing drainage networks, and areas which have recently been affected by continental glaciation generally lack a well-integrated drainage pattern. They are typically swampy and poorly drained. Note the drainage characteristics of the three accompanying topographic maps (Figs. 11-2, 11-3, 11-4).

4. *Eskers.* Small meltwater streams flow in tunnels under a continental glacier. Erosional debris is generally deposited in the stream channels. When the glacier eventually recedes, a small ridge of sediment may remain. These are termed *eskers*. They are readily identified because they are low, narrow, and sinuous ridges which serendipitously wind across a landscape. Goulds Ridge in the Passadumkeag, Maine map (Fig. 11-4) is an example of a partially dissected esker.

5. *Drumlins.* Locally, continental glaciers may override the ground moraine of a previous glacial episode. As they advance, they often shape the older depositional material into elliptical and streamlined hills termed *drumlins*. Drumlins often occur in swarms with their long axis parallel to the direction of ice movement. Most drumlins are asymmetric in their long-axis direction and their gentle slopes commonly face in the direction of ice movement. The Palmyra, New York map area (Fig. 11-3) is an example of a drumlin field.

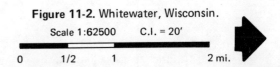

**Figure 11-2.** Whitewater, Wisconsin.

Scale 1:62500    C.I. = 20'

0    1/2    1    2 mi.

**Figure 11-3.** Palmyra, New York.

Scale 1:62500          C.I. = 20'

0                    1                    2 mi.

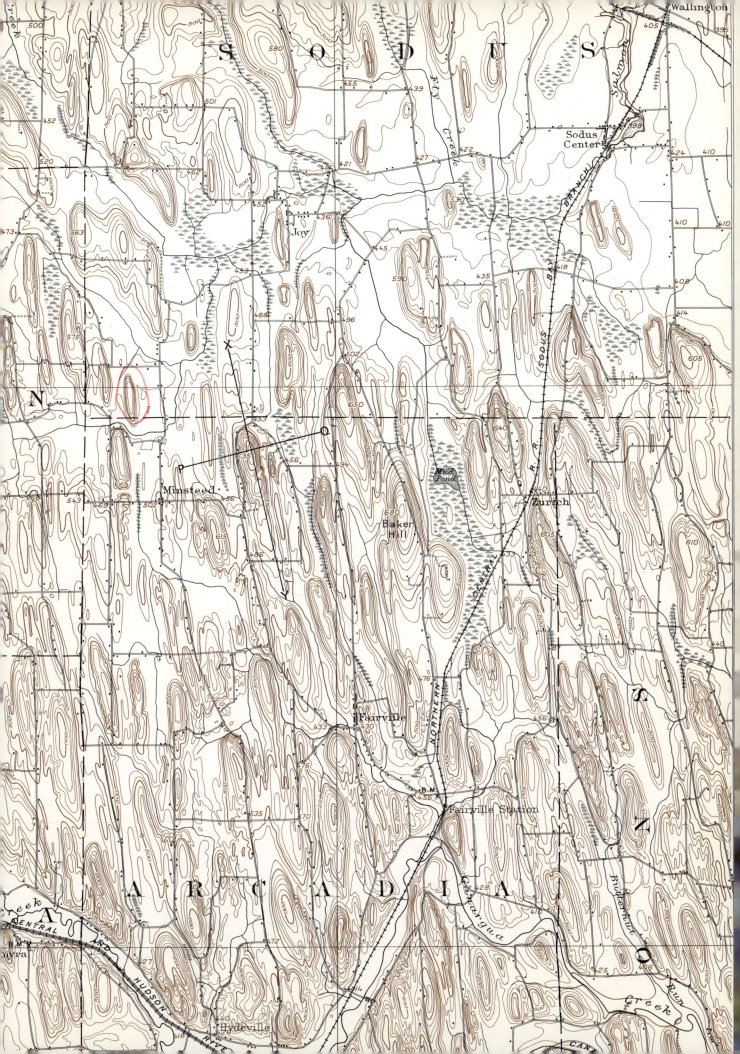

HORSEBACK IS ANOTHER
GROUND MOA

**Figure 11-4.** Passadumkeag, Maine.

Scale 1:62500          C.I. = 20'

0          1/2          1                              2 mi.

102

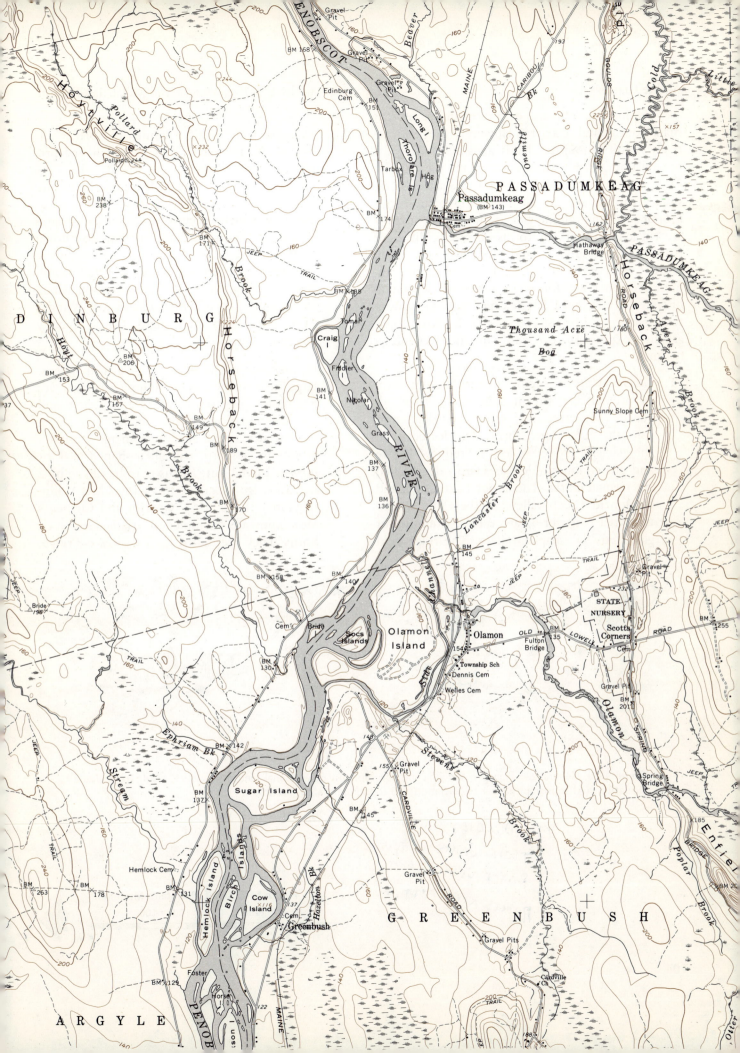

## Alpine Glaciers

Alpine glaciers are of more limited extent than continental glaciers. However, they do actively modify mountain landscapes by eroding the pre-existing stream valleys in which they move. Thus, as opposed to continental glaciation, erosional features are the most important landforms produced during alpine glaciation. In general they are rugged, angular landforms, distinctly different from the more subdued and rounded forms which result form continental glaciation. The more common erosional landforms are listed below and illustrated in Figure 11-5. The Chief Mountain, Montana map (Fig. 11-6) shows the characteristic topographic expression of many of these erosional features.

1. *Glacial Valleys.* Alpine glaciers usually form in pre-existing stream valleys. As they advance, they effectively scour the valleys into smooth, "U"-shaped forms. Divides between the glacially eroded valleys are reduced in size and are commonly left as sharp ridges which are termed *aretes.* The Trout Lake Creek and McDonald Creek valleys in the Chief Mountain area (Fig. 11-6) are typical examples of glaciated "U"-shaped valleys. The ridge held up by Stanton, Vaught and McPartland Mountains is an arete.

2. *Hanging Valleys.* As a result of glacial scouring, small, pre-glacial valleys which had previously merged with larger stream valleys occupied by glaciers, will be left topographically high or "hanging" when the glaciers melt. Streams which occupy these small valleys commonly form waterfalls as they flow down to the new level of the master streams in the glacially deepened valley. In the Chief Mountain area (Fig. 11-6), the small streams which flow southeast off McPartland and Vaught Mountains into McDonald Creek occupy hanging valleys.

3. *Cirques.* At the head of most glacial valleys, large bowl-shaped depressions or *cirques* are common. They represent areas of snow and ice accumulation at the head of an alpine glacier. When several cirques merge around a mountain, a sharp, triangular-shaped peak or *horn* is carved out. Horns may be linked by a narrow ridge called a *col.* In the Chief Mountain area (Fig. 11-6), Oberlin and Clements Mountains are examples of triangular-shaped horns. They are linked by a narrow col ridge.

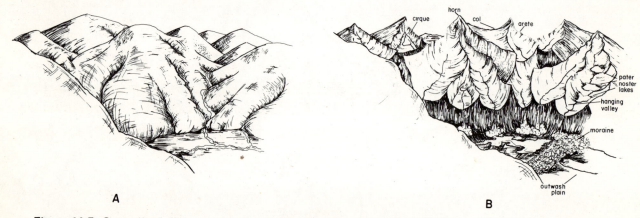

A                    B

**Figure 11-5.** Generalized diagrams showing the development of erosional landforms which result from alpine glaciation: (A) topography prior to glaciation; (B) topography after glaciation.

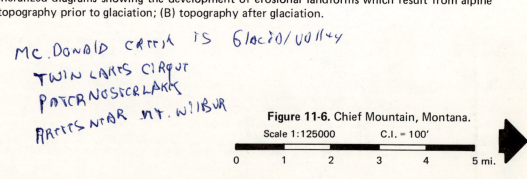

**Figure 11-6.** Chief Mountain, Montana.

Scale 1:125000          C.I. = 100'

0      1      2      3      4      5 mi.

Both peaks are bounded by bowl-shaped cirques.

4. *Glacial Lakes.* Small lakes often form in basins left by glacial scouring. These are called *tarns.* Frequently these occur in a chain down the glacial valley and are then termed *pater noster* lakes. Such a chain of lakes is present in the Chief Mountain area (Fig. 11-6) in the Swiftcurrent stream valley, immediately north of Mt. Grinnell.

Along the terminus of an alpine glacier, erosional debris is deposited and forms landforms similar to those of a continental glacier (although they are usually of smaller scale). These are also termed moraines and outwash plains.

## EXERCISE 11

In this exercise you will examine landforms which typically result from glaciation. These will be contrasted and compared to those which result from the erosive action of streams.

At the end of this exercise you should be able to:

1. Recognize landscapes which have resulted from glacial erosion.
2. Differentiate those which have formed by alpine or continental glaciation.
3. Name characteristic glacial landforms.

### QUESTIONS

1. Whitewater, Wisconsin area (map in manual; Fig. 11-2).

   a. This area has a topography which is typical of regions which have been overridden by continental glaciers. List the most characteristic features which serve to distinguish the landscape from those which result from the erosive action of streams.

   b. On the basis of the distinctive glacial landforms in the area, prepare a generalized paleogeographic map (use a tracing paper overlay) which shows the glacial terminus and other features illustrated in Figure 11-1. From which general direction did the ice advance? What evidence can you use to define this direction?

   c. Is there any topographic evidence which you can use to estimate the approximate position of major meltwater streams which may have flowed outward from the terminus during melting of the glacier? If so, add these streams to your paleogeographic map.

2. Passadumkeag, Maine area (map in manual; Fig. 11-4).

   a. The poorly developed drainage system in this area is characteristic of regions which have recently been affected by continental glaciation. To better visualize the drainage system, prepare a drainage map of the area (use a tracing paper overlay).

   b. What type of drainage pattern is present in this area (refer back to Fig. 8-1)?

   c. Is there a high or low drainage density?

   d. Is the drainage density compatible with the type of sedimentary material which you would expect to find in this area? Discuss your answer.

    e.    Compare your map to those constructed in Exercise 8 and discuss the similarities and differences.

3.  Palmyra, New York area (map in manual; Fig. 11-3).

    a.    This area contains abundant drumlins. To better visualize the shape of drumlins, construct two topographic profiles along the lines shown on the map (use a vertical scale of 1″ = 100′).

    b.    In what general direction did the ice move in the area? Explain your answer.

    c.    Explain the origin of the asymmetrical cross-section of drumlins. How can this asymmetry be used to better define directions of ice movement.

4.  Chief Mountain, Montana area (map in manual; Fig. 11-6).

    a.    The angular and rugged topography in this area typifies that of landscapes formed by alpine glaciation. What characteristics indicate that the topography is not the result of stream erosion?

5.  Eastport and Riverhead, New York 7 1/2′ Quadrangles (supplied in laboratory).

These two maps give complete north-south topographic coverage of a portion of eastern Long Island. Long Island is largely a constructional feature, built of the deposits from two continental glaciers whose southernmost terminations ran roughly parallel to the long axis of the island.

    a.    Using several pieces of graph paper, construct a topographic profile connecting the Long Island shore at Roanoke Landing (EC-NW on the Riverhead map) and the Long Island Country Club (NC-WC on the Eastport map). Use a vertical scale of 1″ = 40′.

    b.    How many topographically distinct areas can you identify on the map (refer to your profile)? Discuss the characteristics of these areas.

    c.    How can you relate these areas to the two periods of continental glaciation?

    d.    Which topographic belts are oldest and which are youngest? On what evidence do you base your conclusions?

    e.    Label your profile with the geomorphic names of the various topographic features.

6.  McCarthy (B-2), Alaska 15′ Quadrangle (supplied in laboratory).

This is an area which is presently being actively eroded by alpine glaciers. Areas of ice are shown in white and have blue contour lines. Note that the glaciers form a system much like that of tributaries and rivers.

    a.    In what general direction is the Klutlan Glacier flowing?

    b.    Note the small glaciers on the south side of the ridge which trends nearly east-west through the central portion of the map. What is the general shape of their northernmost extent? What type of landform would be exposed if the ice melted?

    c.    With reference to the same area as in (b), what is the nature of the area between these small glaciers? What type of landform would be exposed if the ice melted?

    d.    Note the triangular piece of land which protrudes above the glaciers at the eastern end of the ridge described in (b). What type of landform is the glacial erosion producing?

7.  McCarthy (A-7), Alaska 15′ Quadrangle (supplied in laboratory).

This area is near that shown on the previous map, but here the alpine glaciers have almost completely receded. Compare the two maps and attempt to visualize what areas of land exposed in B-2 would have been exposed above the glaciers which once covered the area of A-7.

    a.    Construct a topographic profile along a line connecting Middle Hanagita Lake (northwestern portion of the area) and the 7730′ peak in square 7 (west-central portion of the area). Use a vertical scale of 1″ = 1,000′.

    b.    What is the significance of the shape of the valley in your profile?

    c.    Did the Klu River (west-central portion of the area) cut the valley in which it flows? Explain your answer.

d. Examine the east-west trending ridge in the northwestern portion of the area and compare it to the ridge in the McCarthy B-2 map discussed in parts b, c, and d of the previous question. What would you call the small north-south trending ridges which run normal to this main ridge?

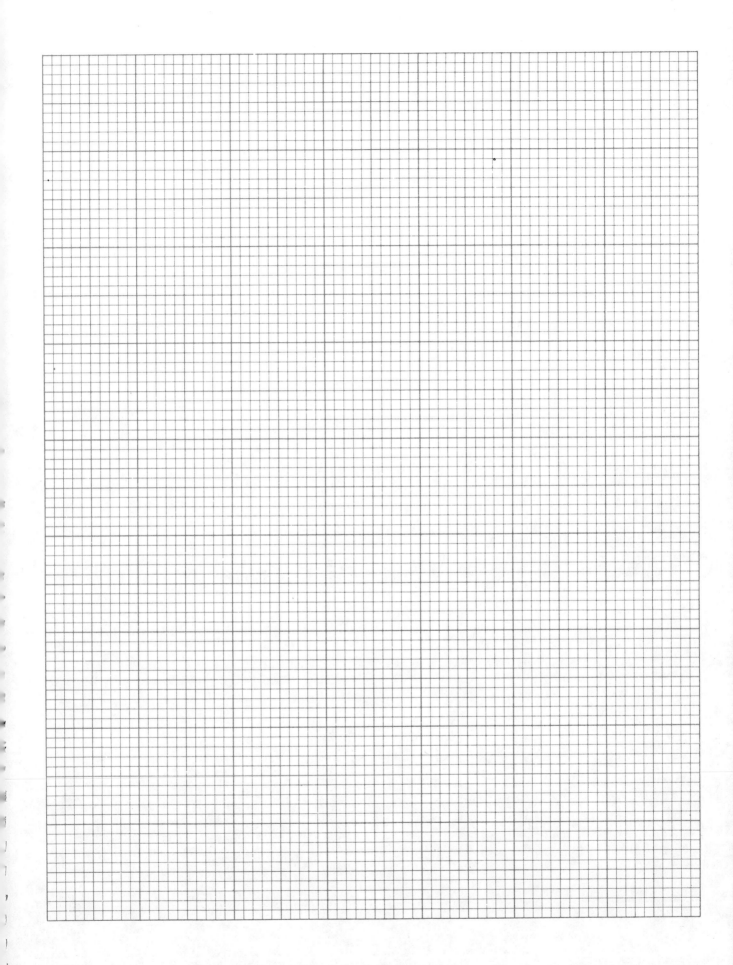

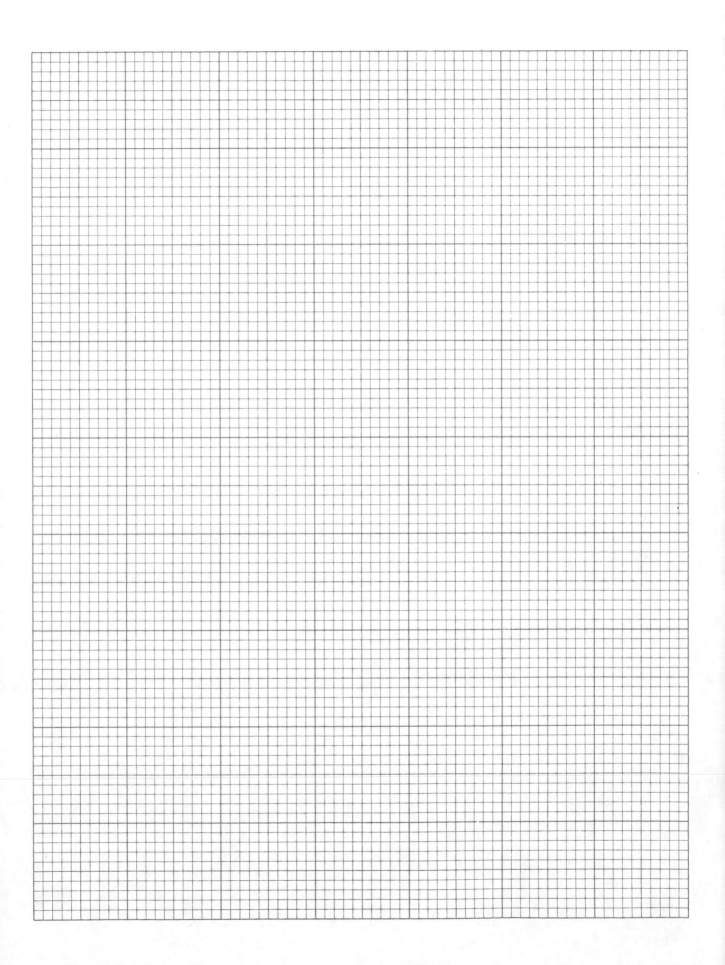

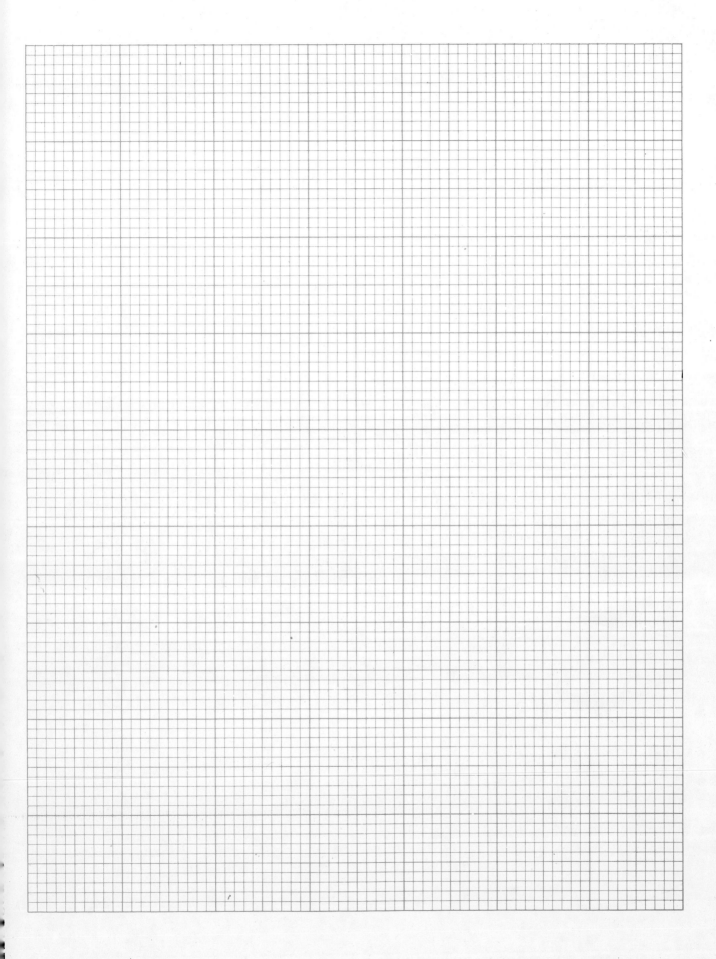

# Shoreline Erosion

Wave and current action along shores of oceans or lakes is an important erosional and depositional agent. However, interpreting the evolutionary geomorphic development of coastal areas is difficult. They commonly have composite landscapes, resulting from the action of waves on inherited landforms which were previously produced by other erosional agents such as streams or glaciers. In addition to this complexity, the relative position of sea level (the erosional base level of shore line processes) commonly fluctuates through geologic time. As a result, shoreline processes are generally not allowed to go through a complete cycle of development. The specific type of coastal landscape which is produced by wave action is thus controlled by two separate factors: (1) the relative position of sea level in the geologic past; and (2) what type of erosional agent was active in producing the landscape which has been affected by wave action. On the basis of the first factor, coastal areas may be classified into two categories: (1) *sub-mergent areas* where sea level has been relatively raised (either by an actual rise in sea level and/or a lowering of the land surface); and (2) *emergent areas* where sea level has been relatively lowered (either by an actual lowering of sea level and/or an elevation of the land surface).

**Submergent Coastlines**

Submergent coastlines are actually drowned landscapes. If the drowned topography was largely produced through stream erosion, a very irregular shoreline will result. An example of a drowned fluvial landscape is the Washington, D.C., Maryland, Virginia area (Fig. 12-1). As glaciated landscapes are typically more streamlined than fluvial terranes, when submerged they are generally represented by straighter, less irregular shorelines. The Boothbay, Maine area typifies a submerged glacial landscape (Fig. 12-2). Large, flooded fluvial or glacial valleys are termed *drowned estuaries.*

Figure 12-1. Washington, D.C., Maryland, Virginia.
Scale 1:250000    C.I. = 50'

0    2    4    6    8    10 mi.

**Figure 12-2.** Boothbay, Maine.

Scale 1:62500     C.I. = 20′

0     1/2     1         2 mi.

The primary effect of wave action in submerged coastal areas is to straighten shoreline irregularities. It consists largely of eroding headlands and depositing the erosional material in embayments. The shoreline landforms which result from these processes are described below and illustrated in Figure 12-3.

Submergent headlands are under active wave erosion. They generally have sharp faces, termed

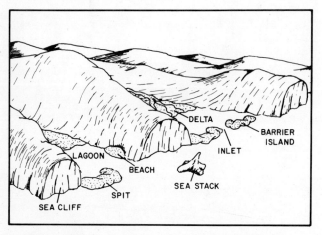

**Figure 12-3.** Erosional and depositional landforms of submergent shorelines.

*sea cliffs,* on the seaward side. Locally, more resistant rocks may be isolated from the mainland and form *sea stacks.* As a result of prevailing incoming wave direction, a current or *longshore drift* commonly parallels most shorelines. This current carries erosional material away from headlands. It may be swept across an embayment and be deposited as a *beach* on the flanks of the next headland. The material may also be deposited as a *spit* connected to the headland from where it was eroded.

Often, spits are progressively enlarged and eventually extend entirely across an embayment, thereby isolating it from open water. Spits may be broken during storms and form *barrier islands.* Behind spits and barrier islands, shallow *lagoons* may develop. Streams which flow from the mainland may build *deltas* and partially fill the lagoons with sediment. If an area remains tectonically stable for a long period of time, these various shoreline processes will combine to produce a linear coastline. The Kingston, Rhode Island area (Fig. 12-4) is a submerged coastal region in which the erosional and depositional action of waves have combined to modify an initally irregular shoreline into a nearly linear one.

**Figure 12-4.** Kingston, Rhode Island.

Scale 1:24000          C.I. = 10′

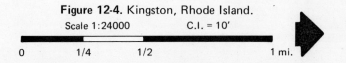

0          1/4          1/2          1 mi.

## Emergent Coastlines

Due to a previous history of wave erosion and deposition, emergent coastlines are generally rather straight. Incoming waves usually break directly against the uplifted coastline and form steep *wave-cut cliffs* which parallel the shore. Most wave energy is directed at the base of these cliffs and often *sea caves* develop. This undercutting eventually leads to slumping and landsliding of the cliff into the sea and causes landward migration of the cliffs. Erosional debris is transported seaward by strong currents and may be deposited as a submarine *wave-built terrace.* Due to the high energy of incoming waves, emergent coastlines typically do not have extensive beaches. Old terraces and beaches which have been elevated above previous sea levels may be found inland from present emergent shorelines. Landforms which are characteristic of an emergent coastline are illustrated in Figure 12-5. The San Pedro, California area (Fig. 12-6) shows the topographic expression of many of these features.

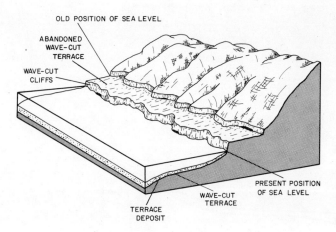

**Figure 12-5.** Erosional and depositional landforms of emergent shorelines.

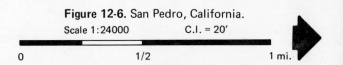

**Figure 12-6.** San Pedro, California.
Scale 1:24000          C.I. = 20'

0                    1/2                    1 mi.

# EXERCISE 12

The characteristic landforms resulting from shoreline erosion will be examined in this exercise. Areas with different histories will be studied and contrasted so that you may gain an understanding of the possible ways in which the topography of coastal areas may develop. At the completion of this laboratory you should be able to:

1.  Distinguish emergent and submergent coastlines.
2.  Name the more important landforms which develop along shorelines.
3.  Describe the geomorphic evolution of a coastal area.

## QUESTIONS

1.  Washington, D.C., Maryland, Virginia area (map in manual; Fig. 12-1).

    a.  Construct a topographic profile along the line shown on the map. Use a vertical scale of 1″ = 100′ (include submarine contours in your profile).

    b.  Indicate on your profile what features have been inherited from a pre-submergence landscape and which have been formed by shoreline erosion and deposition.

    c.  What is a minimum estimate of how much sea level has been relatively raised in this area? What data can you use to make this estimate?

2.  Boothbay, Maine area (map in manual; Fig. 12-2).

    a.  Construct a topographic profile along the line shown in the map. Use a vertical scale of 1″ = 150′ (include submarine contours in your profile).

    b.  How does this profile differ from that of question 1? Discuss how these differences may be used to better define the principal pre-submergence erosional agent.

    c.  Indicate on your profile what features have been inherited from a pre-submergence landscape and which have been formed by shoreline erosion and deposition.

    d.  What do differences in the development of shoreline features in the two areas of questions 1 and 2 indicate about the relative duration of shoreline processes in the two areas?

3.  Kingston, Rhode Island area (map in manual; Fig. 12-4).

    This area has had a complex geomorphic history. The region is underlain by high-grade metamorphic and intrusive igneous rocks. This bedrock terrane was overridden by a continental glacier during the

Pleistocene. The knobby hills in the northwestern portion of the map are part of a terminal moraine. The flat area southwest of the moraine is an outwash plain. Post-glacial stream erosion has dissected the outwash plain. More recent submergence has flooded the combined fluvial-glacial topography.

a. Using a tracing paper overlay, outline the areas which have been inherited from a pre-submergence landscape. Do not include any land which you interpret as having been formed by wave or current deposition.

b. From the appearance of your map, what do you think was the principal erosive agent which produced the pre-submergence landscape? Explain your answer.

c. Point Judith Pond is a drowned estuary. Is it of fluvial or glacial origin? Explain your answer.

d. Are Long, White, and Wash Ponds (NW) related to any coastal processes? Explain their origin.

e. With reference to Figure 12-3, name as many coastal landforms as possible. Add the names to the Kingston map.

f. How has man interrupted the submergent coastline erosional and depositional process?

4. San Pedro, California area (map in manual; Fig. 12-6).

a. Construct a topographic profile along the line shown in the map. Use a vertical scale of 1″ = 800′ (include submarine contours in your profile).

b. Indicate on your profile which features have been formed by shoreline processes and which are inherited.

c. Approximately how much relative drop in sea level has there been in this area? What data can you use to make this estimate?

d. What types of coastal landforms are present along the profile line? (Refer to Fig. 12-5.) Label them in the profile.

5. Toms River, New Jersey 15' Quadrangle (supplied in laboratory).

The narrow strip of land on which Island Beach is located is connected to the mainland north of the map area.

a. Is there any evidence of prevailing longshore drift direction in this area? If so, what is the direction?

b. What geomorphic landform would you call the Island Beach area?

c. Why is the shoreline of the mainland so marshy?

d. What would happen to the marshy areas if Island Beach were destroyed by wave action during a storm?

e. Is this an emergent or submergent coastline? On what criteria do you base your answer?

f. What do you think was the principal erosional agent which produced the landscape which has been affected by shoreline processes in this area? Discuss your answer.

6. San Onofre Bluff, California 7 1/2' Quadrangle (supplied in laboratory).

a. Is this an emergent or submergent coastline? What types of topographic features can be used to answer this question?

b. What type of geomorphic landform is the flat area along which the San Diego Freeway extends?

7. Jekyll Island, Georgia 7 1/2' Quadrangle (supplied in laboratory).

a. Is there evidence to suggest a prevailing longshore drift direction in this area? If so, what is the direction?

b. What type of geomorphic landform is Jekyll Island?

c. Is Jekyll Island an erosional or a depositional landform?

d. What are the low, narrow, elongate ridges which are abundant along the shore of the southeastern tip of Jekyll Island?

e. Is this an emergent or submergent coastal area? On what criteria do you base your answer?

f. How much of the landscape is a result of shoreline processes and how much has been inherited from a previous erosional stage? Discuss your answer.

8. Hyannis, Massachusetts 7 1/2'/8' Quadrangle (supplied in laboratory).

This map covers a north-south portion of Cape Cod. The area was glaciated during the Pleistocene and the east-west trending belt of knobby hills in the central portion of the map area is a terminal moraine. Note the two topographically different shorelines on the north and south sides of the area.

a. What geomorphic feature would you call Sandy Neck (along the north shore)?

b. What geomorphic feature would you call the marshy area south of Sandy Neck?

c. Does the configuration of Sandy Neck suggest a prevailing current or longshore drift direction? If so, what is the direction?

d. Note the breakwater jetties built along the shore of Squaw Island (C-SC). What purpose do the breakwaters serve?

e. How is the Long Beach-Craigville Beach-Squaw Island area related to the Sunset Hill headland?

f. Is there any evidence for a prevailing direction of current or longshore drift in this area? If so, what is the direction?

g. Is this a coastal area of emergence or submergence? On what criteria do you base your answer?


h. Why is the topography of the northern and southern shorelines so different?

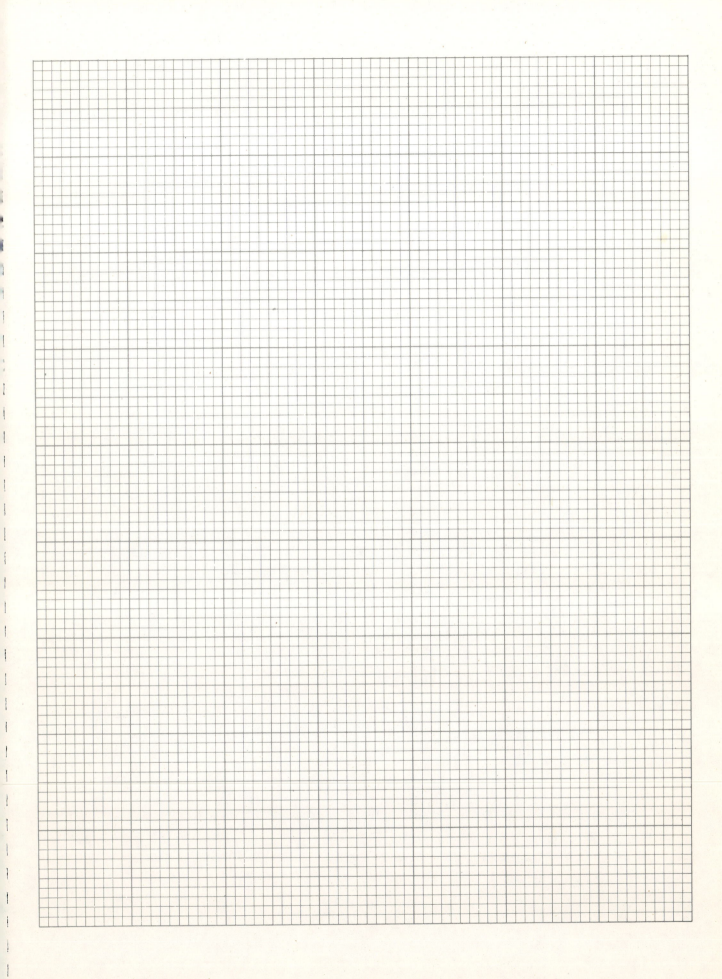

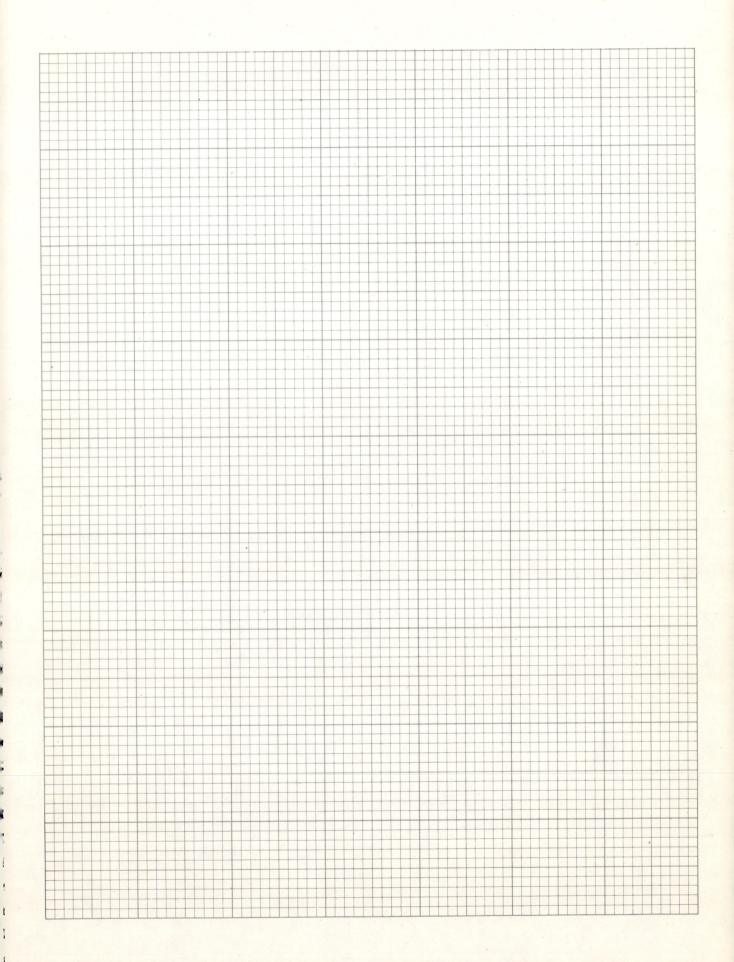

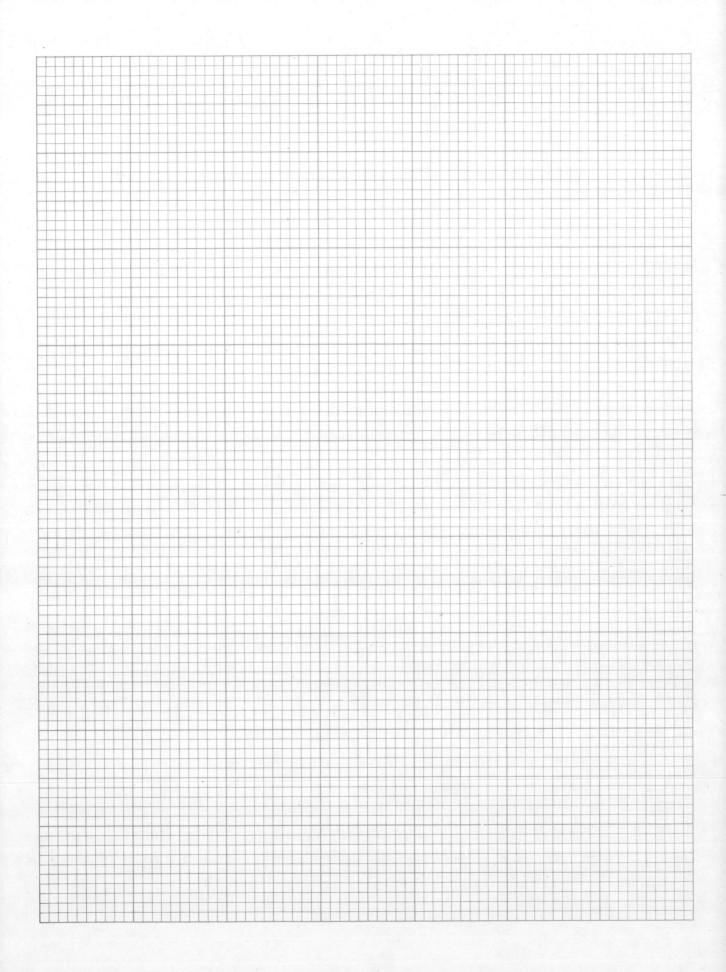

## EXERCISE 13

The following exercise is concerned with describing the three-dimensional attitude of rock units. It is a difficult exercise, as conceptualization of three-dimensional representations takes time. However, with a little patience and help from your laboratory instructor, you will be able to successfully answer the questions. After completion of the exercise you should be able to:

1. Determine the strike and dip of any rock unit by observing its outcrop pattern on a topographic map.
2. Identify and properly label all types of fold structures.
3. Determine the type of relative movement along faults.

### QUESTIONS

1. Study Figure 13-13 and indicate the strike and dip of the various rock units on the top of the figures (use the appropriate symbols and estimate the amount of dip.

   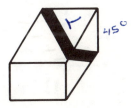

Figure 13-13. Block diagrams illustrating strike and dip.

2. Fill in all blank faces of the block diagrams shown in Figure 13-14 and answer the following questions (A-F refer to the different diagrams). Geographic north is toward the back of each diagram.

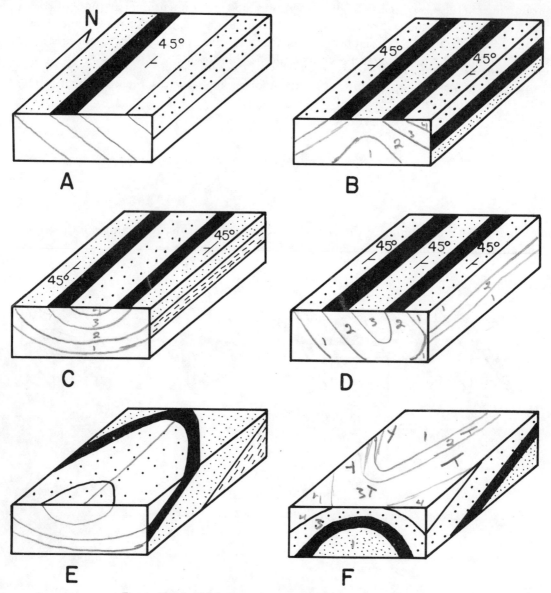

**Figure 13-14.** Block diagrams of various geologic structures.

Diagram A. What is the dip of the rock units (both the general geographic direction and the actual amount of dip in degrees?     E

Diagram B. What kind of structure is this? Label the various rock units with numbers 1, 2, 3, etc. (going from the oldest to the youngest in numerical succession).     ANTI     S

Diagram C. What kind of structure is this? S Y U

Diagram D. What kind of structure is this (fine dotted unit is youngest)? A S T M S Y N C

Diagram E. Fill in the front face and draw the trace of the axial surface on the top. What is the dip of the axial surface if the apparent dip of the beds on the front face is 45°?   45°

148

Diagram F. Sketch a likely interpretation on the top face if the fold is plunging south (toward you). Add several representative dip and strike symbols on the top face.

3. Figure 13-15 is a geologic map (showing the distribution of rock units at the surface of the earth). The land surface is nearly flat except for a slight depression along the stream valley. The ages of the various rock units are indicated by the conventional abbreviated symbols (see the Geologic Time Scale at the front of this manual).

   a. Indicate the strike and dip of the various rock units at several places on the map using the appropriate symbols (see Table 13-1).

   b. What type of structure is present? Accurately draw the trace of the axial surface on the map using the correct symbols.

   c. What happens to the age of the various rock units as the axial trace is approached? get youngra

   d. Draw a cross section along a line connecting the northwestern and southeastern corners of the map (assume a 25° dip for each limb and use the space provided). Include the axial surface in your cross section.

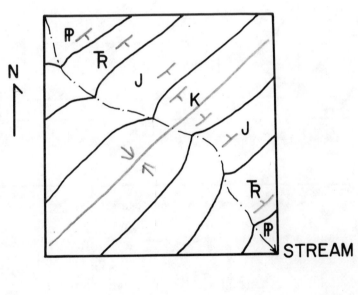

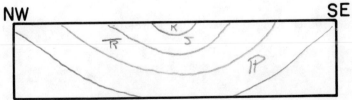

Figure 13-15. Geologic map.

4. Figure 13-16 is a geologic map. The land surface is almost flat except for a small valley along the stream. Ages of the various rock units exposed in this area are indicated by the conventional symbols.

   a. Indicate the strike and dip of the various units at several places on the map.

   b. Draw a cross section along a line connecting the southwestern and northeastern corners of the map (use the space provided). The units dip around 25° in the northeastern half of the area and around 65° in the southwestern half.

   c. What type of structure is present in this area?

   d. Accurately add all necessary structural symbols to the map and to your cross section.

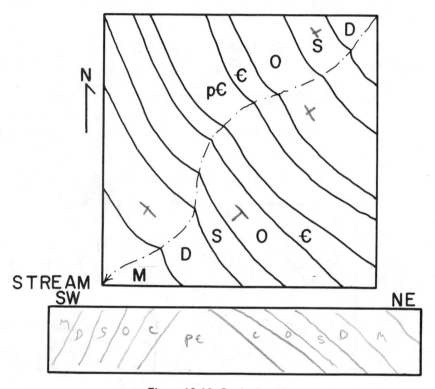

**Figure 13-16.** Geologic map.

5. Two geologic maps of faults are shown in Figure 13-17.
   For Map A.

   a. Which side of the fault is the hanging wall block (upper or lower)?

   b. Did the hanging wall block move relatively up or down?   UP

   c. What type of fault is this?   REVERSE

   For Map B.

   a. Which side of the fault is the footwall block (upper or lower)?

b. Did the footwall block move relatively up or down?  U P

c. What type of fault is this?  NORMAl

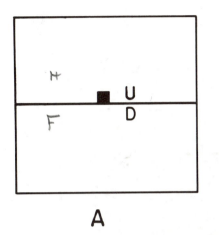

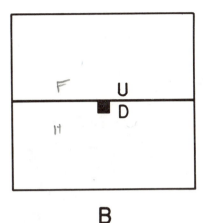

A                                    B

**Figure 13-17.** Geologic maps.

6. The sense of fault movements may also be determined by considering the relative ages of rock units which are offset. In Figure 13-18 geologic maps of two areas are shown. Only three units are mapped (1 is the oldest, 2 intermediate and 3 is the youngest). Strike and dip of the rock units and the fault planes are indicated.

For Map A.

a. Which side of the fault is the hanging wall block (upper or lower)?

b. What type of fault is this? NORMAl Rev

For Map B.

a. Which side of the fault is the footwall block (upper or lower)?

b. Did the hanging wall block move relatively up or down?  UP

c. What type of fault is this?  ReveAST

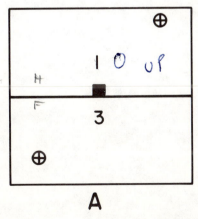

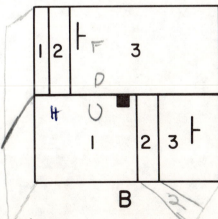

A                                    B

**Figure 13-18.** Geologic maps.

7. Figure 13-19 is a block diagram of a folded area. The fold plunges 30° east. The fold has been cut by a normal fault. Relative ages of units are indicated by conventional symbols.

   a. Carefully complete the front and top of the block diagram.

   b. Add appropriate dip and strike symbols for both the folded units and for the fault surface on the top of the diagram.

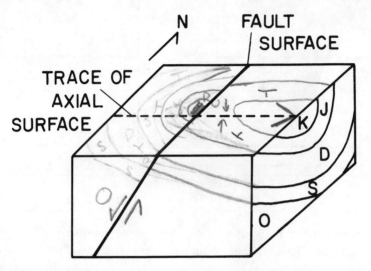

Figure 13-19. Block diagram of a faulted fold.

8. Figure 13-20 is a geologic map of a folded terrain. Topographic relief is negligible except along streams. The relative ages of the various rock units are: A = oldest, D = youngest.

   a. Place appropriate dip and strike symbols in each blank circle on the map.

   b. Complete the contacts between the various rock units.

   c. Add the correct fold symbols (traces of axial surfaces, general plunge directions) to the map.

   d. Color unit A green, B orange, C yellow, D blue.

   e. Draw a generalized geologic cross section along line X-Y (X on left).

   f. Color cross section to correspond with the map.

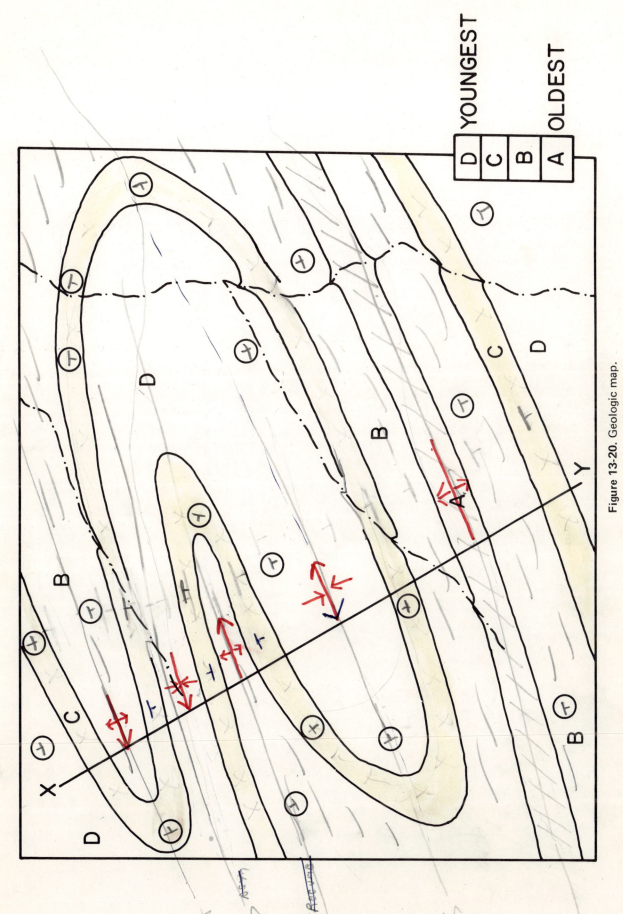

Figure 13-20. Geologic map.

| YOUNGEST | | OLDEST |
|---|---|---|
| D | C | B | A |

# Interpretation of Structural Features on Topographic and Geologic Maps

One of the basic goals of a geologist is to determine the size, shape and areal extent of rock bodies. This requires careful geologic mapping. In the course of this mapping, rocks are divided into groups (commonly called *units* or *formations*) on the basis of similar textures, mineral composition and/or fossil fauna. The boundaries between these units are termed *contacts*. They are traced out on the ground using topographic maps and/or aerial photographs. The result is a geologic map which shows the distribution of rock units at the surface of the earth. Each unit is shown by a different color. Structural measurements made in the field (strike and dip) are shown on the map. With the aid of geologic maps, the structure of an area may be resolved and a geologic cross section constructed which portrays the probable distribution of rock units in the subsurface.

Major structural features such as folds and faults may be recognized by their distinctive patterns on geologic maps. Map patterns resulting from plunging folds have been previously illustrated in Figures 13-6 and 13-7. Contacts of horizontal rock units are also distinctive as they typically parallel topographic contours (Fig. 14-1). Note in Figure 14-1 that horizontal rock units typically have more narrow outcrop patterns on steep slopes than on gentle slopes. Thus, outcrop width on a geologic map is no indication of actual stratigraphic thickness. When rock units are inclined, the outcrop pattern of the deformed units will appear as parallel lines (Fig. 14-2). Variations in outcrop width and the nature of contact "V's" depend on the attitude of rock units and may be used to infer subsurface structure. Angular unconformities may also be recognized on geologic maps as they produce discontinuities in outcrop patterns (Fig. 14-3). Older structures (folds and faults) which are truncated by angular unconformities may appear in erosional depressions in younger rocks which overlie the unconformity (Fig. 14-3). As most faults are nearly linear features which offset rock units, they also usually result in distinctive patterns on geologic maps.

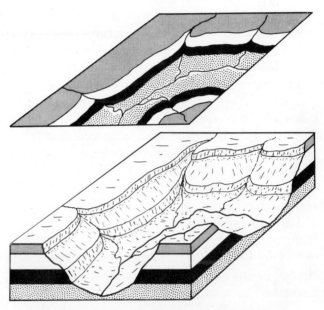

**Figure 14-1.** Block diagram and accompanying geologic map of horizontal rock units.

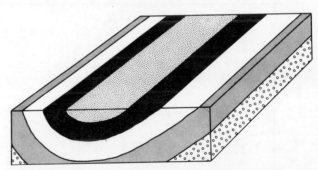

**Figure 14-2.** Generalized block diagram of an asymmetric syncline. Note how outcrop width is a direct function of dip magnitude.

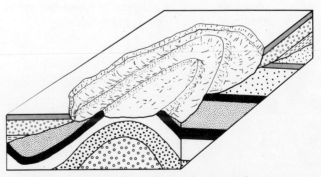

**Figure 14-3.** Block diagram of an angular unconformity.

Although geologic contacts are not shown on topographic maps, they are an invaluable aid in interpretation of geologic structures because the differential erosional resistance of rock units results in topographic expression of fold and fault structures. Areas underlain by horizontal rock units typically have alternating slopes and cliffs whose topographic contours form symmetrical upstream "V's" around drainage features (Fig. 14-1). Inclined units of varying resistance may also be recognized on a topographic map as they often form a series of linear ridges and valleys which parallel strike. Topographic contours on slopes which face in the direction of dip (termed *dip slopes*) are usually regular and more widely spaced than those on slopes which face away from the dip direction (termed *resequent slopes;* Fig. 14-4). When inclined units form the limbs of folds, the direction of dip and strike may be estimated from the topography, and the folds identified as anticlines or synclines. Plunging folds may be recognized on topographic maps because strike ridges will either connect (anticlines) or splay apart (synclines) in the plunge direction.

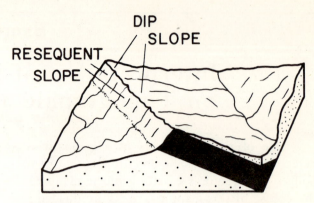

**Figure 14-4.** Block diagram illustrating the relationship of topography to strike and dip of rock units of variable resistance. Sketch a series of contour lines on the diagram.

The topographic expression of a plunging fold is well illustrated in the Tyrone, Pennsylvania region (Fig. 14-5). The area is underlain by a sequence of Paleozoic sedimentary rocks of variable resistance. The stratigraphic section of the southwesternmost portion of the area is listed in Table 14-1. Note that Brush and Canoe Mountains are held up by a very resistant conglomeratic sand-

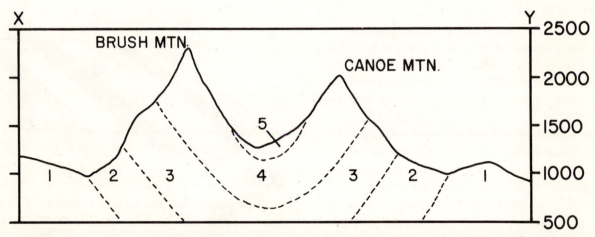

**Figure 14-6.** Topographic profile along line X-Y in the Tyrone, Pennsylvania area (Fig. 14-5); vertical exaggeration = 5.2X. Generalized geologic contacts have been added. Units are numbered as in the stratigraphic column presented in Table 14-1.

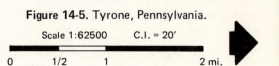

**Figure 14-5.** Tyrone, Pennsylvania.

Scale 1:62500      C.I. = 20'

0      1/2      1      2 mi.

stone unit (the Tuscarora Formation). Generalized geologic contacts between the various units have been added to the accompanying topographic profile (Fig. 14-6). Topographic variation along the profile line indicates that units are dipping toward Canoe Creek on both Brush and Canoe Mountains (compare the profile to Fig. 14-4). This indicates that rocks should become younger in age as Canoe Creek is approached from the two mountains. Inspection of the stratigraphic section confirms this suggestion because Silurian age rocks occur between the mountains and Ordovician rocks occur on their flanks. The topographic map also indicates that Brush and Canoe Mountains connect to the northeast. This suggests that the Tuscarora and other units have been deformed into a plunging fold. Because younger rocks occur between the strike ridges (Brush and Canoe Mountains), the

fold must be a syncline. Synclines plunge away from wraparound units and the fold thus plunges toward the southwest.

Narrow, linear trends which separate topographically different regions are typical of faulted areas. Fault zones may be weakened by movement and thus be unusually susceptible to stream or glacial erosion. Often, small depressions or streams follow the trace of fault planes. In other cases, fault planes are expressed as sharply defined cliffs termed *fault scarps.* Fault controlled topography is characteristic of the Mt. Dome, California-Oregon area (Fig. 14-7). The three north-south trending cliffs in the central portion of the area are fault scarps. The accompanying topographic profile (Fig. 14-8) illustrates how faulting has influenced the topography in this area.

**TABLE 14-1**

Stratigraphic Column for the Southwestern Portion
of the Tyrone, Pennsylvania Area (Fig. 14-5).

Silurian
5. Clinton Formation—shale and interbedded siltstone.
4. Tuscarora Formation—conglomeratic sandstone.
3. Juniata Formation—sandstone and minor interbedded shale.

Ordovician
2. Bald Eagle Formation—sandstone and interbedded shale.
1. Reedsville and Curtin Formations—shale and minor interbedded sandstone at top, limestone and minor interbedded shale at bottom.

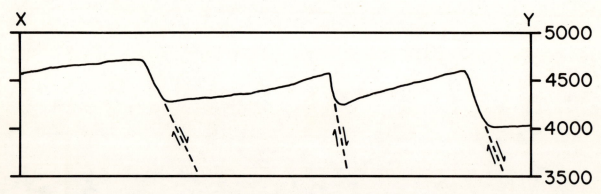

**Figure 14-8.** Topographic profile along line X-Y in the Mt. Dome, California-Oregon area (Fig. 14-7); vertical exaggeration = 5.2X. Generalized geologic structure indicated.

**Figure 14-7.** Mt. Dome, California-Oregon.

Scale 1:62500      C.I. = 40'

0      1/2      1                                    2 mi.

## EXERCISE 14

In this exercise various types of geologic structures will be examined on topographic and geologic maps. Projection of geologic structures in the subsurface will be achieved through construction of geologic cross sections.

After completion of the exercise you should be able to:

1. Recognize folds and faults on geologic and topographic maps.
2. Construct geologic cross sections and infer subsurface relations.

### QUESTIONS

1. Geologic map of the Bright Angel Quadrangle, Arizona (supplied in laboratory).

    This is perhaps the most famous geologic map ever made. It is unusual both for the excellent exposure afforded by the steep walls of the Grand Canyon and for the completeness of the stratigraphic section exposed.

    a.   What is the attitude of the Redwall Limestone?

    b.   What is the range of absolute age (in millions of years) represented by the stratigraphic section exposed in this area?

    c.   How many angular unconformities are present in this area?

    d.   What is the approximate thickness of the Cambrian section exposed in this area (in feet)?

    e.   Why is Bright Angel Canyon more nearly linear than the Colorado River Valley?

2. Tyrone, Pennsylvania area (map in manual; Fig. 14-5).

    a.   Figure 14-6 presents a topographic profile along line X-Y to illustrate how topography may be used to infer the dip of fold limbs. Construct a topographic profile along line P-Q using the same vertical exaggeration as in Figure 14-6. What relationship between topography and plunge direction can you define?

    b.   Using Figure 14-6 as a guide, add generalized geologic contacts to your profile.

    c.   On a tracing paper overlay, prepare a drainage map for the area. What type of drainage pattern is present?

    d.   Using the stratigraphic column presented in Table 14-1 and the generalized geologic contacts shown in Figure 14-6, prepare a geologic map for the Brush and Canoe Mountain area. Pay close attention to dip direction and stream gradients in tracing geologic contacts. Color the geologic map according to the following scheme: Clinton Formation—blue, Tuscarora Formation—red, Juniata Formation—orange, Bald Eagle Formation—green, Reedsville and Curtin Formations—yellow.

    e.   Add representative strike and dip symbols to your geologic map. Also add appropriate symbols to indicate axial surface trace and plunge direction orientations for the fold.

161

f. Discuss the relationship between the structure of this area and the drainage system.

g. Both Brush and Canoe Mountains are held up by the same unit. However, they have different elevations (Fig. 14-6). How can you explain this anomaly?

3. Point Reyes, California 19'/15' Quadrangle (supplied in laboratory).

a. What major geologic structures can you identify in this area?

b. How are they topographically expressed?

4. Summerville, Georgia 7 1/2' Quadrangle (supplied in laboratory).
This area is underlain by deformed Paleozoic rocks of variable resistance.

a. Using several sheets of graph paper, construct a topographic profile along a line connecting Marble Spring (EC-WC) and the intersection of an intermittant stream (near trail) and the southern border of the map (SW-SE). Use a vertical scale of 1" = 150'.

b. What type of fold is defined by Taylor Ridge?

c. What is the strike of the trace of its axial surface?

d. What type of fold is defined by Taylor Ridge and Simms Mountain?

e. In what direction does it plunge?

f. Is the fold symmetrical or asymmetrical? Explain your answer.

5. Geologic map of the Stickleyville, Virginia 7 1/2' Quadrangle (supplied in laboratory).

a. Construct a topographic profile along a line connecting the small road at the northern edge of the map (where two trails split off; NC-NC) and Miles Cemetary (SE-NC). Use a vertical scale of 1" = 100'.

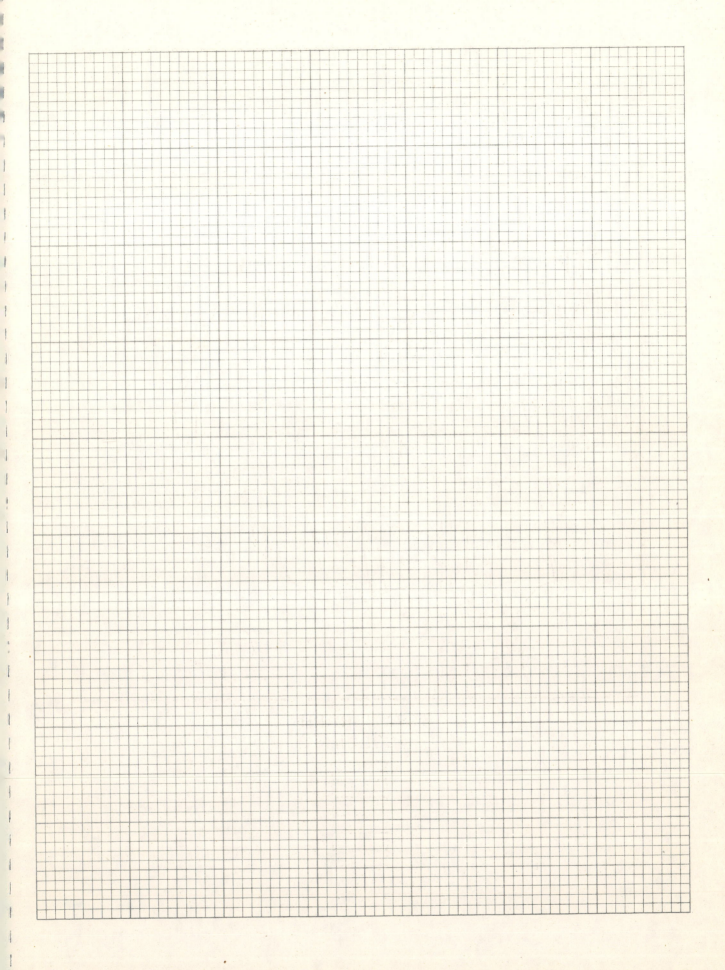

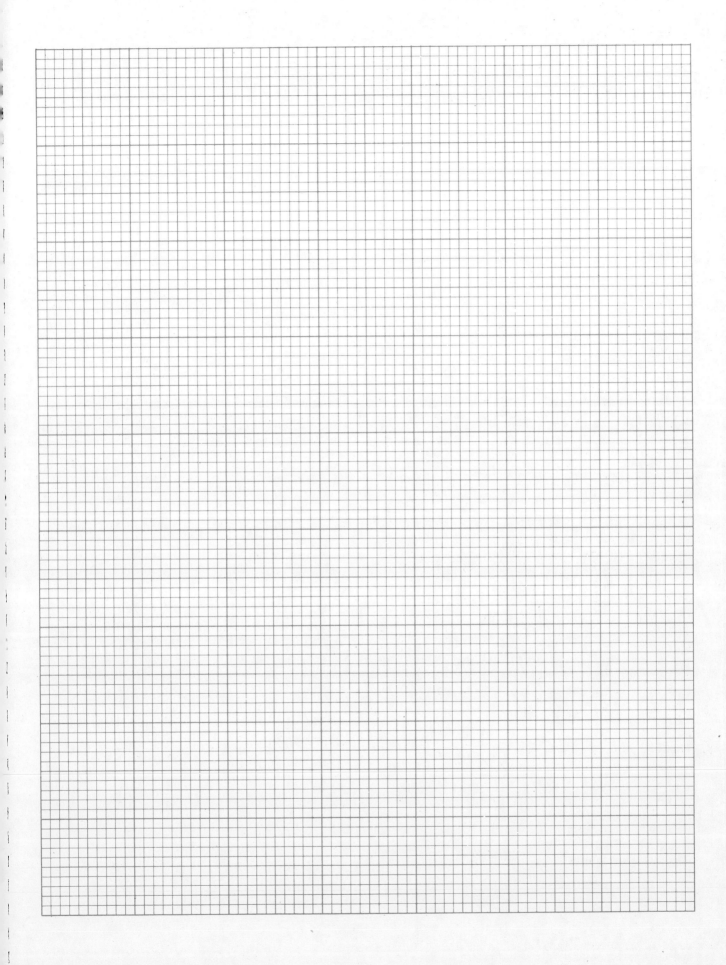

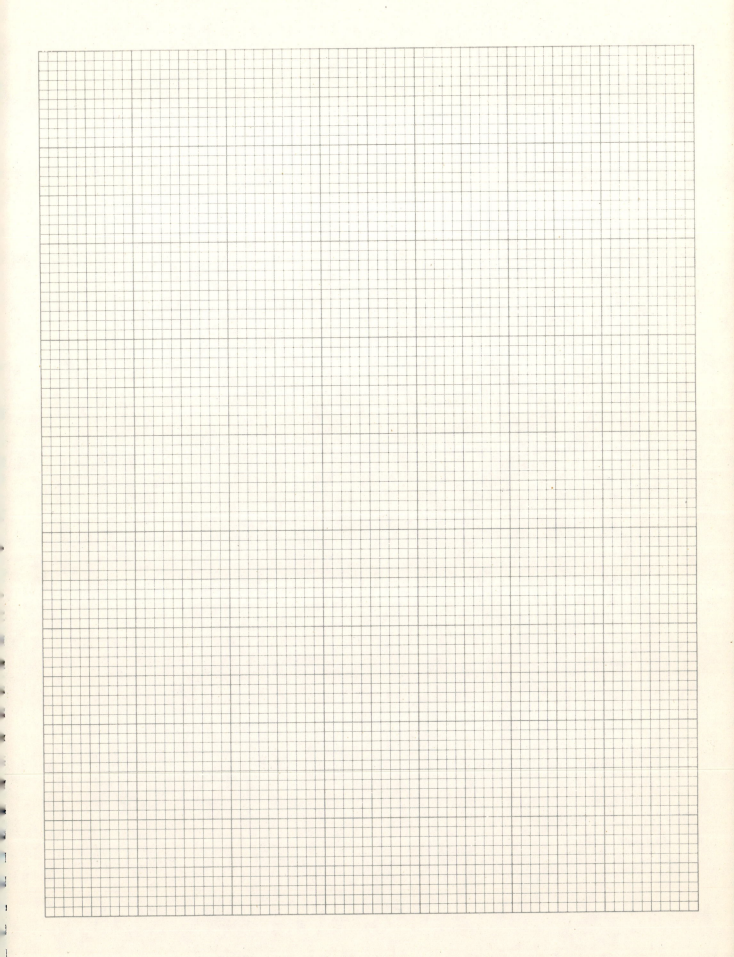

# CLASS NOTES

# CLASS NOTES